Department of The Treasury Study on the Sporting Suitability of Modified Semiautomatic Assault Rifles

April 1998

TABLE OF CONTENTS

Page

EXECUTIVE SUMMARY

On November 14, 1997, the President and the Secretary of the Treasury ordered a review of the importation of certain modified versions of semiautomatic assault rifles into the United States.[1] The decision to conduct this review stemmed in part from concerns expressed by members of Congress and others that the rifles being imported were essentially the same as semiautomatic assault rifles previously determined to be nonimportable in a 1989 decision by the Bureau of Alcohol, Tobacco and Firearms (ATF). The decision also stemmed from the fact that nearly 10 years had passed since the last comprehensive review of the importation of rifles, and many new rifles had been developed during this time.

Under 18 U.S.C. section 925(d)(3), the Secretary shall approve applications for importation only when the firearms are generally recognized as particularly suitable for or readily adaptable to sporting purposes (the "sporting purposes test"). In 1989, ATF denied applications to import a series of semiautomatic versions of automatic-fire military assault rifles. When ATF examined these semiautomatic assault rifles, it found that the rifles, while no longer machineguns, still had a military configuration that was designed for killing and disabling the enemy and that distinguished the rifles from traditional sporting rifles. This distinctively military configuration served as the basis for ATF's finding that the rifles were not considered sporting rifles under the statute.

The military configuration identified by ATF incorporated eight physical features: ability to accept a detachable magazine, folding/telescoping stocks, separate pistol grips, ability to accept a bayonet, flash suppressors, bipods, grenade launchers, and night sights. In 1989, ATF took the position that any of these military configuration features, other than the ability to accept a detachable magazine, would make a semiautomatic rifle not importable.

Subsequent to the 1989 decision, certain semiautomatic assault rifles that failed the 1989 sporting purposes test were modified to remove all of the military configuration features other than the ability to accept a detachable magazine. Significantly, most of these modified rifles not only still had the ability to accept a detachable magazine but, more specifically, still had the ability to accept a detachable large capacity magazine that

---

[1]   The President and the Secretary directed that all pending and future applications for importation of these rifles not be acted upon until completion of the review. They also ordered that outstanding permits for importation of the rifles be suspended for the duration of the review period. The existence of applications to import 1 million new rifles and outstanding permits for nearly 600,000 other rifles threatened to defeat the purpose of the expedited review unless the Department of the Treasury deferred action on additional applications and temporarily suspended the outstanding permits. (See exhibit 1 for a copy of the November 14, 1997, memorandum directing this review.)

The rifles that are the subject of this review are referred to in this report as "study rifles."

was originally designed and produced for the military assault rifles from which they were derived. These magazines are referred to in this report as "large capacity military magazines." Study rifles with the ability to accept such magazines are referred to in this report as "large capacity military magazine rifles," or "LCMM rifles." It appears that only one study rifle, the VEPR caliber .308 (an AK47 variant), is not an LCMM rifle. Based on the standard developed in 1989, these modified rifles were found to meet the sporting purposes test. Accordingly, the study rifles were approved for import into the United States.

These modified rifles are the subject of the present review. Like the rifles banned in 1989, the study rifles are semiautomatic rifles based on AK47, FN-FAL, HK91 and 93, Uzi, and SIG SG550 military assault rifles. While there are at least 59 specific model designations of the study rifles, they all fall within the basic designs listed above. There are at least 39 models based on the AK47 design, 8 on the FN-FAL design, 7 on the HK91 and 93 designs, 3 on the Uzi design, and 2 on the SIG SG550 design (see exhibit 2 for a list of the models). Illustrations of some of the study rifles are included in exhibit 3 of this report.

This review takes another look at the entire matter to determine whether the modified rifles approved for importation since 1989 are generally recognized as particularly suitable for or readily adaptable to sporting purposes.[2] We have explored the statutory history of the sporting purposes test and prior administrative and judicial interpretations; reexamined the basic tenets of the 1989 decision; analyzed the physical features of the study rifles, as well as information from a wide variety of sources relating to the rifles' use and suitability for sporting purposes; and assessed changes in law that might have bearing on the treatment of the rifles.

This review has led us to conclude that the basic finding of the 1989 decision remains valid and that military-style semiautomatic rifles are not importable under the sporting purposes standard. Accordingly, we believe that the Department of the Treasury correctly has been denying the importation of rifles that had any of the distinctly military configuration features identified in 1989, other than the ability to accept a detachable magazine. Our review, however, did result in a finding that the ability to accept a detachable large capacity magazine originally designed and produced for a military assault weapon should be added to the list of disqualifying military configuration features identified in 1989.

Several important changes have occurred since 1989 that have led us to reevaluate the importance of this feature in the sporting purposes test. Most significantly, by passing the 1994 bans on semiautomatic assault weapons and large capacity ammunition feeding

---

[2] The study was carried out by a working group composed of ATF and Treasury representatives. The working group's activities and findings were overseen by a steering committee composed of ATF and Treasury officials.

devices, Congress sent a strong signal that firearms with the ability to expel large amounts of ammunition quickly are not sporting; rather, firearms with this ability have military purposes and are a crime problem. Specifically, Congress found that these magazines served "combat-functional ends" and were attractive to criminals because they "make it possible to fire a large number of rounds without reloading, then to reload quickly when those rounds are spent."[3] Moreover, we did not find any evidence that the ability to accept a detachable large capacity military magazine serves any sporting purpose. Accordingly, we found that the ability to accept such a magazine is a critical factor in the sporting purposes test, which must be given the same weight as the other military configuration features identified in 1989.

In addition, the information we collected on the use and suitability of LCMM rifles for hunting and organized competitive target shooting demonstrated that the rifles are not especially suitable for sporting purposes. Although our review of this information indicated that, with certain exceptions, the LCMM rifles sometimes are used for hunting, their actual use in hunting is limited. There are even some general restrictions and prohibitions on the use of semiautomatic rifles for hunting game. Similarly, although the LCMM rifles usually may be used, with certain exceptions, and sometimes are used for organized competitive target shooting, their suitability for this activity is limited. In fact, there are some restrictions and prohibitions on their use.

Furthermore, the information we gathered demonstrated that the LCMM rifles are attractive to certain criminals. We identified specific examples of the LCMM rifles' being used in violent crime and gun trafficking. In addition, we found some disturbing trends involving the LCMM rifles, including a rapid and continuing increase in crime gun trace requests after 1991 and a rapid "time to crime." Their ability to accept large capacity military magazines likely plays a role in their appeal to these criminals.

After weighing all the information collected, we found that the LCMM rifles are not generally recognized as particularly suitable for or readily adaptable to sporting purposes and are therefore not importable. However, this decision will in no way preclude the importation of true sporting firearms.

---

[3]   H. Rep. No. 103-489, at 18-19.

BACKGROUND

Importation of Firearms Under the Gun Control Act

The Gun Control Act of 1968 (GCA)[4] generally prohibits the importation of firearms into the United States.[5] However, the GCA creates four narrow categories of firearms that the Secretary of the Treasury shall authorize for importation. The category that is relevant to this study is found at 18 U.S.C. section 925(d)(3).

> The Secretary shall authorize a firearm . . . to be imported or brought into the United States . . . if the firearm . . .
>
> > (3) is of a type that does not fall within the definition of a firearm as defined in section 5845(a) of the Internal Revenue Code of 1954 and **is generally recognized as particularly suitable for or readily adaptable to sporting purposes**, excluding surplus military firearms, except in any case where the Secretary has not authorized the importation of the firearm pursuant to this paragraph, it shall be unlawful to import any frame, receiver, or barrel of such firearm which would be prohibited if assembled. (Emphasis added)

This provision originally was enacted, in a slightly different form, by Title IV of the Omnibus Crime Control and Safe Streets Act of 1968[6] and also was contained in Title I of the GCA, which amended Title IV later that year.

The GCA was enacted in large part "to assist law enforcement authorities in the States and their subdivisions in combating the increasing prevalence of crime in the United States." However, the Senate Report to the act also made clear that Congress did not intend the GCA to place any undue or unnecessary restrictions or burdens on responsible, law-abiding citizens with respect to acquiring, possessing, transporting, or using firearms for lawful activities.[7]

---

[4] Pub. L. No. 90-618.

[5] 18 U.S.C. section 922(l).

[6] Pub. L. No. 90-351.

[7] S. Rep. No. 1501, 90th Cong. 2d Sess. 22 (1968).

Consistent with this general approach, legislative history indicates that Congress intended the importation standard provided in section 925(d)(3) to exclude military-type weapons from importation to prevent such weapons from being used in crime, while allowing the importation of high-quality sporting rifles. According to the Senate Report, section 925(d)(3) was intended to "curb the flow of surplus military weapons and other firearms being brought into the United States which are not particularly suitable for target shooting or hunting."[8] The report goes on to explain that "[t]he importation of certain foreign-made and military surplus nonsporting firearms has an important bearing on the problem which this title is designed to alleviate [crime]. Thus, the import provisions of this title seem entirely justified."[9] Indeed, during debate on the bill, Senator Dodd, the sponsor of the legislation, stated that "Title IV prohibits importation of arms which the Secretary determines are not suitable for . . . sport . . . . The entire intent of the importation section is to get those kinds of weapons that are used by criminals and have no sporting purpose."[10]

The Senate Report, however, also makes it clear that the importation standards "are designed and intended to provide for the importation of quality made, sporting firearms, including . . . rifles such as those manufactured and imported by Browning and other such manufacturers and importers of firearms."[11] (The rifles being imported by Browning at that time were semiautomatic and manually operated traditional sporting rifles of high quality.) Similarly, the report states that the importation prohibition "would not interfere with the bringing in of currently produced firearms, such as rifles . . . of recognized quality which are used for hunting and for recreational purposes."[12] The reference to recreational purposes is not inconsistent with the expressed purpose of restricting importation to firearms particularly suitable for target shooting or hunting, because firearms particularly suitable for these purposes also can be used for other purposes such as recreational shooting.

During debate on the bill, there was discussion about the meaning of the term "sporting purposes." Senator Dodd stated:

> [h]ere again I would have to say that if a military weapon is used in a

---

[8] S. Rep. No. 1501, 90th Cong. 2d Sess. 22 (1968).

[9] S. Rep. No. 1501, 90th Cong. 2d Sess. 24 (1968).

[10] 114 Cong. Rec. S 5556, 5582, 5585 (1968).

[11] S. Rep. No. 1501, 90th Cong. 2d. Sess. 38 (1968).

[12] S. Rep. No. 1501, 90th Cong. 2d. Sess. 22 (1968).

special sporting event, it does not become a sporting weapon. It is a
military weapon used in a special sporting event . . . . As I said previously
the language says no firearms will be admitted into this country unless they
are genuine sporting weapons.[13]

Legislative history also shows that the determination of a weapon's suitability for sporting
purposes is the direct responsibility of the Secretary of the Treasury. The Secretary was
given this discretion largely because Congress recognized that section 925(d)(3) was a
difficult provision to implement. Immediately after discussing the large role cheap
imported .22 caliber revolvers were playing in crime, the Senate Report stated:

> [t]he difficulty of defining weapons characteristics to meet this target
> without discriminating against sporting quality firearms, was a major
> reason why the Secretary of the Treasury has been given fairly broad
> discretion in defining and administering the import prohibition.[14]

Indeed, Congress granted this discretion to the Secretary even though some expressed
concern with its breadth:

> [t]he proposed import restrictions of Title IV would give the Secretary of
> the Treasury unusually broad discretion to decide whether a particular type
> of firearm is generally recognized as particularly suitable for, or readily
> adaptable to, sporting purposes. If this authority means anything, it
> permits Federal officials to differ with the judgment of sportsmen expressed
> through consumer preference in the marketplace . . . . [15]

Section 925(d)(3) provides that the Secretary shall authorize the importation of a firearm
if it is of a "type" that is generally recognized as particularly suitable for or readily
adaptable to sporting purposes. The legislative history also makes it clear that the
Secretary shall scrutinize types of firearms in exercising his authority under section 925(d).
Specifically, the Senate Report to the GCA states that section 925(d) "gives the

Secretary authority to permit the importation of ammunition and certain types of
firearms."[16]

---

[13]   114 Cong. Rec. 27461-462 (1968).

[14]   S. Rep. No. 1501, 90th Cong. 2d Sess. 38 (1968).

[15]   S. Rep. No. 1097, 90th Cong. 2d. Sess. 2155 (1968) (views of Senators Dirksen, Hruska, Thurmond, and
       Burdick). In Gun South, Inc. v. Brady, F.2d 858, 863 (11th Cir. 1989), the court, based on legislative
       history, found that the GCA gives the Secretary "unusually broad discretion in applying section 925(d)(3)."

[16]   S. Rep. No. 1501, 90th Cong. 2d. Sess. 38 (1968).

The Senate Report to the GCA also recommended that the Secretary establish a council that would provide him with guidance and assistance in determining which firearms meet the criteria for importation into the United States.[17] Accordingly, following the enactment of the GCA, the Secretary established the Firearms Evaluation Panel (FEP) (also known as the Firearms Advisory Panel) to provide guidelines for implementation of the "sporting purposes" test. This panel was composed of representatives from the military, the law enforcement community, and the firearms industry. At the initial meeting of the FEP, it was understood that the panel's role would be advisory only.[18] The panel focused its attention on handguns and recommended the adoption of factoring criteria to evaluate the various types of handguns. These factoring criteria are based upon such considerations as overall length of the firearm, caliber, safety features, and frame construction. ATF thereafter developed an evaluation sheet (ATF Form 4590) that was put into use for evaluating handguns pursuant to section 925(d)(3). (See exhibit 4.)

The FEP did not propose criteria for evaluating rifles and shotguns under section 925(d)(3). Other than surplus military firearms, which Congress addressed separately, the rifles and shotguns being imported prior to 1968 were generally conventional rifles and shotguns specifically intended for sporting purposes. Therefore, in 1968, there was no cause to develop criteria for evaluating the sporting purposes of rifles and shotguns.

1984 Application of the Sporting Purposes Test

The first time that ATF undertook a meaningful analysis of rifles or shotguns under the sporting purposes test was in 1984. At that time, ATF was faced with a new breed of imported shotgun, and it became clear that the historical assumption that all shotguns were sporting was no longer viable. Specifically, ATF was asked to determine whether the Striker-12 shotgun was suitable for sporting purposes. This shotgun is a military/law enforcement weapon initially designed and manufactured in South Africa for riot control. When the importer was asked to submit evidence of the weapon's sporting purposes, it provided information that the weapon was suitable for police/combat-style competitions. ATF determined that this type of competition did not constitute a sporting purpose

under the statute, and that the shotgun was not suitable for the traditional shotgun sports of hunting, and trap and skeet shooting.

---

[17] S. Rep. No. 1501, 90th Cong. 2d Sess. 38 (1968).

[18] Gilbert Equipment Co. v. Higgins, 709 F. Supp. 1071, 1083, n. 7 (S.D. Ala. 1989), aff'd without op., 894 F.2d 412 (11th Cir. 1990).

1986 Firearms Owners Protection Act

On May 19, 1986, Congress passed the Firearms Owners Protection Act,[19] which amended section 925(d)(3) to provide that the Secretary "shall" (instead of "may") authorize the importation of a firearm that is of a type that is generally recognized as particularly suitable for or readily adaptable to sporting purposes. The Senate Report to the law stated "it is anticipated that in the vast majority of cases, [the substitution of 'shall' for 'may' in the authorization section] will not result in any change in current practices."[20] As the courts have found, "[r]egardless of the changes made [by the 1986 law], the firearm must meet the sporting purposes test and it remains the Secretary's obligation to determine whether specific firearms satisfy this test."[21]

1986 Application of the Sporting Purposes Test

In 1986, ATF again had to determine whether a shotgun met the sporting purposes test, when the Gilbert Equipment Company requested that the USAS-12 shotgun be classified as a sporting firearm under section 925(d)(3). Again, ATF refused to recognize police/combat-style competitions as a sporting purpose. After examining and testing the weapon, ATF determined its weight, size, bulk, designed magazine capacity, configuration, and other factors prevented it from being classified as particularly suitable for or readily adaptable to the traditional shotgun sports of hunting, and trap and skeet shooting. Accordingly, its importation was denied.

When this decision was challenged in Federal court, ATF argued, in part, that large magazine capacity and rapid reloading ability are military features. The court accepted this argument, finding "the overall appearance and design of the weapon (especially the detachable box magazine . . .) is that of a combat weapon and not a sporting weapon."[22] In reaching this decision, the court was not persuaded by the importer's argument that box magazines can be lengthened or shortened depending on desired shell capacity.[23] The court also agreed with ATF's conclusion that police/combat-style competitions were not considered sporting purposes.

---

[19] Pub. L. No. 99-308.

[20] S. Rep. No. 98-583, 98th Cong. 1st Sess. 27 (1984).

[21] Gilbert Equipment Co., 709 F. Supp. at 1083.

[22] Id. at 1089.

[23] Id. at 1087, n. 20 and 1089.

1989 Report on the Importability of Semiautomatic Assault Rifles

In 1989, after five children were killed in a California schoolyard by a gunman with a semiautomatic copy of an AK47, ATF decided to reexamine whether certain semiautomatic assault-type rifles met the sporting purposes test. This decision was reached after consultation with the Director of the Office of National Drug Control Policy. In March and April 1989, ATF announced that it was suspending the importation of certain "assault-type rifles." For the purposes of this suspension, assault-type rifles were those rifles that generally met the following criteria: (1) military appearance; (2) large magazine capacity; and (3) semiautomatic version of a machinegun. An ATF working group was established to reevaluate the importability of these assault-type rifles. On July 6, 1989, the group issued its Report and Recommendation of the ATF Working Group on the Importability of Certain Semiautomatic Rifles (hereinafter 1989 report).

In the 1989 report, the working group first discussed whether the assault-type rifles under review fell within a "type" of firearm for the purposes of section 925(d)(3). The working group concluded that most of the assault-type rifles under review represented "a distinctive type of rifle [which it called the "semiautomatic assault rifle"] distinguished by certain general characteristics which are common to the modern military assault rifle."[24] The working group explained that the modern military assault rifle is a weapon designed for killing or disabling the enemy and has characteristics designed to accomplish this purpose. Moreover, it found that these characteristics distinguish modern military assault rifles from traditional sporting rifles.

The characteristics of the modern military assault rifle that the working group identified were as follows: (1) military configuration (which included: ability to accept a detachable magazine, folding/telescoping stocks, separate pistol grips, ability to accept a bayonet, flash suppressors, bipods, grenade launchers, and night sights) (see exhibit 5 for a thorough discussion of each of these features); (2) ability to fire automatically (i.e., as a machinegun); and (3) chambered to accept a centerfire cartridge case having a length of 2.25 inches or less.[25] In regards to the ability to accept a detachable magazine, the working group explained that:

> [v]irtually all modern military firearms are designed to accept large, detachable magazines. This provides the soldier with a fairly large ammunition supply and the ability to rapidly reload. Thus, large capacity magazines are indicative of military firearms. While detachable

---

[24] 1989 report at 6.

[25] 1989 report at 6.

magazines are not limited to military firearms, most traditional
semiautomatic sporting firearms, designed to accommodate a detachable
magazine, have a relatively small magazine capacity.[26]

The working group emphasized that these characteristics had to be looked at as a whole to
determine whether the overall configuration of each of the assault-type rifles under review
placed the rifle fairly within the semiautomatic assault rifle type. The semiautomatic
assault rifles shared all the above military assault rifle characteristics other than being
machineguns.[27]

The working group also addressed the scope of the term "sporting purposes." It
concluded that the term should be given a narrow interpretation that focuses on the
traditional sports of hunting and organized competitive target shooting. The working
group made this determination by looking to the statute, its legislative history, applicable
case law, the work of the FEP, and prior interpretations by ATF. In addition, the working
group found that the reference to sporting purposes was intended to stand in contrast to
military and law enforcement applications. Consequently, it determined that
police/combat-type competitions should not be treated as sporting activities.[28]

The working group then evaluated whether the semiautomatic assault rifle type of firearm
is generally recognized as particularly suitable for or readily adaptable to traditional
sporting applications. This examination took into account technical and marketing data,
expert opinions, the recommended uses of the firearms, and information on the actual uses
for which the weapons are employed in this country. The working group, however, did
not consider criminal use as a factor in its analysis of the importability of this type of
firearm.

After analyzing this information, the working group concluded that semiautomatic assault
rifles are not a type of firearm generally recognized as particularly suitable for or readily
adaptable to sporting purposes. Accordingly, the working group concluded that semiautomatic assault rifles should not be authorized for importation under section 925(d)(3).
However, the working group found that some of the assault-type rifles under review (the
Valmet Hunter and .22 rimfire caliber rifles), did not fall within the semiautomatic assault
rifle type. In the case of the Valmet Hunter, the working group found that although it was
based on the operating mechanism of the AK47 assault rifle, it had been substantially

---

[26]   1989 report at 6 (footnote omitted).

[27]   The semiautomatic assault rifles were semiautomatic versions of machineguns.

[28]   1989 report at 9-11.

changed so that it was similar to a traditional sporting rifle.[29] Specifically, it did not have any of the military configuration features identified by the working group, except for the ability to accept a detachable magazine.

Following the 1989 study, ATF took the position that a semiautomatic rifle with any of the eight military configuration features identified in the 1989 report, other than the ability to accept a detachable magazine, failed the sporting purposes test and, therefore, was not importable.

Gun South, Inc. v. Brady

Concurrent with its work on the 1989 report, ATF was involved in litigation with Gun South, Inc. (GSI). In October 1988 and February 1989, ATF had granted GSI permits to import AUG-SA rifles. As mentioned previously, in March and April of 1989, ATF imposed a temporary suspension on the importation of rifles being reviewed in the 1989 study, which included the AUG-SA rifle. GSI filed suit in Federal court, seeking to prohibit the Government from interfering with the delivery of firearms imported under permits issued prior to the temporary suspension.

The court of appeals found that the Government had the authority to suspend temporarily the importation of GSI's AUG-SA rifles because the GCA "impliedly authorizes" such action.[30] In addition, the court rejected GSI's contention that the suspension was arbitrary and capricious because the AUG-SA rifle had not physically changed, explaining the argument "places too much emphasis on the rifle's structure for determining whether a firearm falls within the sporting purpose exception. While the Bureau must consider the rifle's physical structure, the [GCA] requires the Bureau to equally consider the rifle's use."[31] In addition, the court found that ATF adequately had considered sufficient evidence before imposing the temporary suspension, citing evidence ATF had considered

demonstrating that semiautomatic assault-type rifles were being used with increasing frequency in crime.[32]

---

[29] This finding reflects the fact that the operating mechanism of the AK-47 assault rifle is similar to the operating mechanism used in many traditional sporting rifles.

[30] Gun South, Inc. v. Brady, 877 F.2d 858 (11th Cir. 1989). The court of appeals issued its ruling just days before the 1989 report was issued. However, the report was complete before the ruling was issued.

[31] Id.

[32] Id.

Although GSI sued ATF on the temporary suspension of its import permits, once the 1989 report was issued, no one pursued a lawsuit challenging ATF's determination that the semiautomatic assault rifles banned from importation did not meet the sporting purposes test.[33]

Violent Crime Control and Law Enforcement Act of 1994

On September 13, 1994, Congress passed the Violent Crime Control and Law Enforcement Act of 1994,[34] which made it unlawful, with certain exceptions, to manufacture, transfer, or possess semiautomatic assault weapons as defined by the statute.[35] The statute defined semiautomatic assault weapons to include 19 named models of firearms (or copies or duplicates of the firearms in any caliber);[36] semiauto-matic rifles that have the ability to accept detachable magazines and have at least two of five features specified in the law; semiautomatic pistols that have the ability to accept detachable magazines and have at least two of five features specified in the law; and semiautomatic shotguns that have at least two of four features specified in the law.[37] However, Congress

---

[33] After the 1989 report was issued, Mitchell Arms, Inc. asserted takings claims against the Government based upon the suspension and revocation of four permits allowing for the importation of semiautomatic assault rifles and ATF's temporary moratorium on import permits for other rifles. The court found for the Government, holding the injury complained of was not redressable as a taking because Mitchell Arms did not hold a property interest within the meaning of the Just Compensation Clause of the Fifth Amendment. Mitchell Arms v. United States, 26 Cl. Ct. 1 (1992), aff'd, 7 F.3d 212 (Fed. Cir. 1993), cert. denied, 511 U.S. 1106 (1994).

[34] Pub. L. No. 103-22. Title XI, Subtitle A of this act may be cited as the "Public Safety and Recreational Firearms Use Protection Act."

[35] 18 U.S.C. section 922(v).

[36] Chapter 18 U.S.C. section 921(a)(30)(A) states that the term "semiautomatic assault weapon" means "any of the firearms, or copies or duplicates of the firearms in any caliber, known as -," followed by a list of named firearms. Even though section 921(a)(3) defines "firearm" as used in chapter 18 to mean, in part, "the frame or receiver of any such weapon," the use of "firearm" in section 921(a)(30)(A) has not been interpreted to mean a frame or receiver of any of the named weapons, except when the frame or receiver actually is incorporated in one of the named weapons.

Any other interpretation would be contrary to Congress' intent in enacting the assault weapon ban. In the House Report to the assault weapon ban, Congress emphasized that the ban was to be interpreted narrowly. For example, the report explained that the present bill was more tightly focused than earlier drafts which gave ATF authority to ban any weapon which "embodies the same configuration" as the named list of guns in section 921(a)(30)(A); instead, the present bill "contains a set of specific characteristics that must be present in order to ban any additional semiautomatic assault weapons [beyond the listed weapons]." H. Rep. 103-489 at 21.

[37] 18 U.S.C. section 921(a)(30).

exempted from the assault weapon ban any semiautomatic rifle that cannot accept a detachable magazine that holds more than five rounds of ammunition and any semiautomatic shotgun that cannot hold more than five rounds of ammunition in a fixed or detachable magazine.[38]

Although the 1994 law was not directly addressing the sporting purposes test in section 925(d)(3), section 925(d)(3) had a strong influence on the law's content. The technical work of ATF's 1989 report was, to a large extent, incorporated into the 1994 law. The House Report to the 1994 law explained that although the legal question of whether semiautomatic assault weapons met section 925(d)(3)'s sporting purposes test "is not directly posed by [the 1994 law], the working group's research and analysis on assault weapons is relevant on the questions of the purposes underlying the design of assault weapons, the characteristics that distinguish them from sporting guns, and the reasons underlying each of the distinguishing features."[39] As in the 1989 study, Congress focused on the external features of firearms, rather than on their semiautomatic operating mechanism.

The 1994 law also made it unlawful to possess and transfer large capacity ammunition feeding devices manufactured after September 13, 1994.[40] A large capacity ammunition feeding device was generally defined as a magazine, belt, drum, feed strip, or similar device that has the capacity of, or that can be readily restored or converted to accept, more than 10 rounds of ammunition.[41]

Congress passed these provisions of the 1994 law in response to the use of semiautomatic assault weapons and large capacity ammunition feeding devices in crime. Congress had been presented with much evidence demonstrating that these weapons were "the weapons of choice among drug dealers, criminal gangs, hate groups, and mentally deranged persons bent on mass murder."[42] The House Report to the 1994 law recounts numerous crimes that had occurred involving semiautomatic assault weapons and large capacity magazines that were originally designed and produced for military assault rifles.[43]

---

[38] 18 U.S.C. sections 922(v)(3)(C)&(D).

[39] II. Rep. No. 103-489, at 17, n. 19.

[40] 18 U.S.C. section 922(w).

[41] 18 U.S.C. section 921(a)(31).

[42] H. Rep. No. 103-489, at 13.

[43] H. Rep. No. 103-489, at 14-15.

In enacting the semiautomatic assault weapon and large capacity ammunition feeding device bans, Congress emphasized that it was not preventing the possession of sporting firearms. The House Report, for example, stated that the bill differed from earlier bills in that "it is designed to be more tightly focused and more carefully crafted to clearly exempt legitimate sporting guns."[44] In addition, Congress specifically exempted 661 long guns from the assault weapon ban which are "most commonly used in hunting and recreational sports."[45]

Both the 1994 law and its legislative history demonstrate that Congress recognized that ammunition capacity is a factor in determining whether a firearm is a sporting firearm. For example, large capacity ammunition feeding devices were banned, while rifles and shotguns with small ammunition capacities were exempted from the assault weapon ban. Moreover, the House Report specifically states that the ability to accept a large capacity magazine was a military configuration feature which was not "merely cosmetic," but "serve[d] specific, combat-functional ends."[46] The House Report also explains that, while "[m]ost of the weapons covered by the [ban] come equipped with magazines that hold 30 rounds [and can be replaced with magazines that hold 50 or even 100 rounds], . . . [i]n contrast, hunting rifles and shotguns typically have much smaller magazine capabilities-- from 3-5."[47]

Finally, it must be emphasized that the semiautomatic assault weapon ban of section 922(v) is distinct from the sporting purposes test governing imports of section 925(d)(3). Clearly, any weapon banned under section 922(v) cannot be imported into the United States because its possession in the United States would be illegal. However, it is possible that a weapon not defined as a semiautomatic assault weapon under section 922(v) still would not be importable under section 925(d)(3). In order to be importable, the firearm must be of a type generally recognized as particularly suitable for or readily adaptable to sporting purposes regardless of its categorization under section 922(v). The

Secretary's discretion under section 925(d)(3) remains intact for all weapons not banned by the 1994 statute.

The Present Review

Prior to the November 14, 1997, decision to conduct this review, certain members of

---

[44] H. Rep. No. 103-489, at 21.

[45] H. Rep. No. 103-489, at 20. None of these 661 guns are study rifles.

[46] H. Rep. No. 103-489, at 18.

[47] H. Rep. No. 103-489, at 19 (footnote omitted).

Congress strongly urged that it was necessary to review the manner in which the Treasury Department is applying the sporting purposes test to the study rifles, in order to ensure that the present practice is consistent with section 925(d)(3) and current patterns of gun use. The fact that it had been nearly 10 years since the last comprehensive review of the importation of rifles (with many new rifles being developed during this time) also contributed to the decision to conduct this review.

DEFINING THE TYPE OF WEAPON UNDER REVIEW

Section 925 (d) (3) provides that the Secretary shall authorize the importation of a firearm if it is of a "type" that meets the sporting purposes test. Given this statutory mandate, we had to determine whether the study rifles suspended from importation fell within one type of firearm. Our review of the study rifles demonstrated that all were derived from semiautomatic assault rifles that failed to meet the sporting purposes test in 1989 but were later found to be importable when certain military features were removed.

Within this group, we determined that virtually all of the study rifles shared another important feature: The ability to accept a detachable large capacity magazine (e.g., more than 10 rounds) that was originally designed and produced for one of the following military assault rifles: AK47, FN-FAL, HK91 or 93, SIG SG550, or Uzi. (This is the only military configuration feature cited in the 1989 study that remains with any of the study rifles).

We determined that all of the study rifles that shared both of these characteristics fell within a type of firearm which, for the purposes of this report, we call "large capacity military magazine rifles" or "LCMM rifles." It appears that only one study rifle, the VEPR caliber .308--which is based on the AK47 design--does not fall within this type because it does not have the ability to accept a large capacity military magazine.

SCOPE OF "SPORTING PURPOSES"

As in the 1989 study, we had to determine the scope of "sporting purposes" as used in section 925(d)(3). Looking to the statute, its legislative history, the work of the Firearms Evaluation Panel (see exhibit 6), and prior ATF interpretations, we determined sporting purposes should be given a narrow reading, incorporating only the traditional sports of hunting and organized competitive target shooting (rather than a broader interpretation that could include virtually any lawful activity or competition.)

In terms of the statute itself, the structure of the importation provisions suggests a somewhat narrow interpretation. Firearms are prohibited from importation (section 922(l)), with four specific exceptions (section 925(d)). A broad interpretation permitting a firearm to be imported because someone may wish to use it in some lawful shooting activity would render the general prohibition of section 922(l) meaningless.

Similarly, as discussed in the "Background" section, the legislative history of the GCA indicates that the term sporting purposes narrowly refers to the traditional sports of hunting and organized competitive target shooting. There is nothing in the history to indicate that it was intended to recognize every conceivable type of activity or competition that might employ a firearm.

In addition, the FEP specifically addressed the informal shooting activity of "plinking" (shooting at randomly selected targets such as bottles and cans) and determined that it was not a legitimate sporting purpose under the statute. The panel found that, "while many persons participate in this type of activity and much ammunition was expended in such endeavors, it was primarily a pastime and could not be considered a sport for the purposes of importation. . . ." (See exhibit 6.)

Finally, the 1989 report determined that the term sporting purposes should be given a narrow reading incorporating the traditional rifle sports of hunting and organized competitive target shooting. In addition, the report determined that the statute's reference to sporting purposes was intended to stand in contrast with military and law enforcement applications. This is consistent with ATF's interpretation in the context of the Striker-12 shotgun and the USAS-12 shotgun. It is also supported by the court's decision in Gilbert Equipment Co. v. Higgins.

We received some comments urging us to find "practical shooting" is a sport for the purposes of section 925(d)(3).[48] Further, we received information showing that practical shooting is gaining in popularity in the United States and is governed by an organization that has sponsored national events since 1989. It also has an international organization.

While some may consider practical shooting a sport, by its very nature it is closer to police/combat-style competition and is not comparable to the more traditional types of sports, such as hunting and organized competitive target shooting. Therefore, we are not convinced that practical shooting does, in fact, constitute a sporting purpose under section 925(d)(3).[49] However, even if we were to assume for the sake of argument that practical shooting is a sport for the purposes of the statute, we still would have to decide whether a firearm that could be used in practical shooting meets the sporting purposes test. In other words, it still would need to be determined whether the firearm is of a type that is generally recognized as particularly suitable for or readily adaptable to practical shooting and other sporting purposes.[50] Moreover, the legislative history makes clear that the use of a military weapon in a practical shooting competition would not make that weapon

---

[48] Practical shooting involves moving, identifying, and engaging multiple targets and delivering a number of shots rapidly. In doing this, practical shooting participants test their defensive skills as they encounter props, including walls and barricades, with full or partial targets, "no-shoots," steel reaction targets, movers, and others to challenge them.

[49] As noted earlier, ATF has taken the position that police/combat-style competitions do not constitute a "sporting purpose." This position was upheld in Gilbert Equipment Co., 709 F. Supp. at 1077.

[50] Our findings on the use and suitability of the LCMM rifles in practical shooting competitions are contained in the "Suitability for Sporting Purposes" section of this report.

sporting: "if a military weapon is used in a special sporting event, it does not become a sporting weapon. It is a military weapon used in a special sporting event."[51] While none of the LCMM rifles are military weapons, they still retain the military feature of the ability to accept a large capacity military magazine.

[51]   114 Cong. Rec. 27461-462 (1968) (Sen. Dodd).

METHOD OF STUDY

As explained in the "Executive Summary" section of this report, the purpose of this study is to review whether modified semiautomatic assault rifles are properly importable under 18 U.S.C. section 925(d)(3). More specifically, we reexamined the conclusions of the 1989 report as applied today to determine whether we are correct to allow importation of the study rifles that have been modified by having certain military features removed. To determine whether such rifles are generally recognized as particularly suitable for or readily adaptable to sporting purposes, the Secretary must consider both the physical features of the rifles and the actual uses of the rifles.[52] Because it appears that all of the study rifles that have been imported to date have the ability to accept a large capacity military magazine,[53] all of the information collected on the study rifles' physical features and actual uses applies only to the LCMM rifles.

Physical features:

The discussion of the LCMM rifles' physical features are contained in the "Suitability for Sporting Purposes" section of this report.

Use:

We collected relevant information on the use of the LCMM rifles. Although the 1989 study did not consider the criminal use of firearms in its importability analysis, legislative history demonstrates and the courts have found that criminal use is a factor that can be considered in determining whether a firearm meets the requirements of section 925(d)(3).[54] Accordingly, we decided to consider the criminal use of the LCMM rifles in the present analysis.

The term "generally recognized" in section 925(d)(3) indicates that the Secretary should base his evaluation of whether a firearm is of a type that is particularly suitable for or readily adaptable to sporting purposes, in part, on a "community standard" of the firearm's use.[55] The community standard "may change over time even though the firearm remains the same. Thus, a changing pattern of use may significantly affect whether a firearm is generally recognized as particularly suitable for or readily adaptable to a sporting purpose."[56] Therefore, to assist the Secretary in determining whether the LCMM rifles presently are of a type generally recognized as particularly suitable for or readily adaptable to sporting purposes, we gathered information from the relevant "community." The relevant community was defined as persons and groups who are

---

[52] Gun South, Inc., 877 F.2d at 866.

[53] The VEPR caliber .308 discussed on page 16 has not yet been imported.

[54] 114 Cong. Rec. S 5556, 5582, 5585 (1968)("[t]he entire intent of the importation section [of the sporting purposes test] is to get those kinds of weapons that are used by criminals and have no sporting purposes") (Sen. Dodd); Gun South, Inc., 877 F.2d at 866.

[55] Gun South, Inc., 877 F.2d at 866.

[56] Id.

knowledgeable about the uses of these firearms or have relevant information about whether these firearms are particularly suitable for sporting purposes. We identified more than 2,000 persons or groups we believed would be able to provide relevant, factual information on these issues. The individuals and groups were selected to obtain a broad range of perspectives on the issues. We conducted surveys to obtain specific information from hunting guides, editors of hunting and shooting magazines, organized competitive shooting groups, State game commissions, and law enforcement agencies and organizations. Additionally, we asked industry members, trade associations, and various interest and information groups to provide relevant information.[57] A detailed presentation of the surveys and responses is included as an appendix to this report.

We also reviewed numerous advertisements and publications, both those submitted by the editors of hunting and shooting magazines and those collected internally, in our search for material discussing the uses of the LCMM rifles. Further, we collected importation data, tracing data, and case studies.[58]

Our findings on use are contained in the "Suitability for Sporting Purposes" section of this report.

---

[57]   **Hunting guides**: Guides were asked about specific types of firearms used by their clients. The guides were an easily definable group, versus the entire universe of hunters. We obtained the names of the hunting guides surveyed from the States.

Editors of hunting and shooting magazines: Editors were surveyed to determine whether they recommended the LCMM rifles for hunting or organized competitive target shooting and whether they had written any articles on the subject. The list of editors we surveyed was obtained from a directory of firearms-related organizations.

Organized competitive shooting groups: Organized groups were asked whether they sponsored competitive events with high-power semiautomatic rifles and whether the LCMM rifles were allowed in those competitions. We felt it was significant to query those who are involved with organized events rather than unofficial activities with no specific rules or guidelines. As with the editors above, the list of groups was obtained from a directory of firearms-related organizations.

State game commissions: State officials were surveyed to determine whether the use of the LCMM rifles was prohibited or restricted for hunting in each State.

Law enforcement agencies and organizations: Specific national organizations and a sampling of 26 police departments across the country were contacted about their knowledge of the LCMM rifles' use in crime. The national organizations were surveyed with the intent that they would gather input from the wide range of law enforcement agencies that they represent or that they would have access to national studies on the subject.

Industry members and trade associations: These groups were included because of their knowledge on the issue.

Interest and information groups: These organizations were included because of their wide range of perspectives on the issue.

[58]   To assist us with our review of the crime-related information we collected, we obtained the services of Garen J. Wintemute, MD, M.P.H. Director of the Violence Prevention Research Program, University of California, Davis, and Anthony A. Braga, Ph.D., J.F.K. School of Government, Harvard University.

SUITABILITY FOR SPORTING PURPOSES

The next step in our review was to evaluate whether the LCMM rifles, as a type, are generally recognized as particularly suitable for or readily adaptable to hunting and organized competitive target shooting.[59] The standard applied in making this determination is high. It requires more than a showing that the LCMM rifles may be used or even are sometimes used for hunting and organized competitive target shooting; if this were the standard, the statute would be meaningless. Rather, the standard requires a showing that the LCMM rifles are especially suitable for use in hunting and organized competitive target shooting.

As discussed in the "Method of Study" section, we considered both the physical features of the LCMM rifles and the actual uses of the LCMM rifles in making this determination.

<u>Physical Features</u>

The ability to accept a detachable large capacity magazine that was originally designed and produced for one of the following military assault rifles: AK47, FN-FAL, HK91 or 93, SIG SG550, or Uzi.

Although the LCMM rifles have been stripped of many of their military features, they all still have the ability to accept a detachable large capacity magazine that was originally designed and produced for one of the following military assault rifles: AK47, FN-FAL, HK91 and 93, SIG SG550, or Uzi; in other words, they still have a feature that was designed for killing or disabling an enemy. As the 1989 report explains:

> Virtually all modern military firearms are designed to accept large, detachable magazines. This provides the soldier with a fairly large ammunition supply and the ability to rapidly reload. Thus, large capacity magazines are indicative of military firearms. While detachable magazines are not limited to military firearms, most traditional

---

[59] One commenter suggests that the Secretary has been improperly applying the "readily adaptable to sporting purposes" provision of the statute. Historically, the Secretary has considered the "particularly suitable for or readily adaptable to" provisions as one standard. The broader interpretation urged by the commenter would make the standard virtually unenforceable. If the Secretary allowed the importation of a firearm which is readily adaptable to sporting purposes, without requiring it actually to be adapted prior to importation, the Secretary would have no control over whether the adaptation actually would occur following the importation.

semiautomatic sporting firearms, designed to accommodate a detachable
magazine, have a relatively small magazine capacity.[60]

Thus, the 1989 report found the ability to accept a detachable large capacity magazine
originally designed and produced for a military assault rifle was a military, not a sporting,
feature. Nevertheless, in 1989 it was decided that the ability to accept such a large
capacity magazine, in the absence of other military configuration features, would not be
viewed as disqualifying for the purposes of the sporting purposes test. However, several
important developments, which are discussed below, have led us to reevaluate the weight
that should be given to the ability to accept a detachable large capacity military magazine
in the sporting purposes test.

Most significantly, we must reevaluate the significance of this military feature because of a
major amendment that was made to the GCA since the 1989 report was issued. In 1994,
as discussed in the "Background" section of this report, Congress passed a ban on large
capacity ammunition feeding devices and semiautomatic assault weapons.[61] In enacting
these bans, Congress made it clear that it was not preventing the possession of sporting
firearms.[62] Although the 1994 law was not directly addressing the sporting purposes test,
section 925(d)(3) had a strong influence on the law's content. As discussed previously,
the technical work of ATF's 1989 report was, to a large extent, incorporated into the 1994
law.

Both the 1994 law and its legislative history demonstrate that Congress found that
ammunition capacity is a factor in whether a firearm is a sporting firearm. For example,
large capacity ammunition feeding devices were banned, while rifles and shotguns with
small ammunition capacities were exempted from the assault weapon ban. In other words,
Congress found magazine capacity to be such an important factor that a semiautomatic
rifle that cannot accept a detachable magazine that holds more than five rounds of
ammunition will not be banned, even if it contains all five of the assault

---

[60] 1989 report at 6 (footnote omitted). This was not the first time that ATF considered magazine capacity to
be a relevant factor in deciding whether a firearm met the sporting purposes test. See Gilbert Equipment
Co., 709 F. Supp. at 1089 ("the overall appearance and design of the weapon (especially the detachable box
magazine . . .) is that of a combat weapon and not a sporting weapon."

[61] The ban on large capacity ammunition feeding devices does not include any such device manufactured on
or before September 13, 1994. Accordingly, there are vast numbers of large capacity magazines originally
designed and produced for military assault weapons that are legal to transfer and possess ("grandfathered"
large capacity military magazines). Presently these grandfathered large capacity military magazines fit the
LCMM rifles.

[62] See, for example, H. Rep. No. 103-489, at 21.

weapon features listed in the law. Moreover, unlike the assault weapon ban in which a detachable magazine and at least two physical features are required to ban a rifle, a large capacity magazine in and of itself is banned.

In addition, the House Report specifically states that the ability to accept a large capacity magazine is a military configuration characteristic that is not "merely cosmetic," but "serve[s] specific, combat-functional ends."[63]   The House Report also explains that large capacity magazines

> make it possible to fire a large number of rounds without re-loading, then to reload quickly when those rounds are spent. Most of the weapons covered by the proposed legislation come equipped with magazines that hold 30 rounds. Even these magazines, however, can be replaced with magazines that hold 50 or even 100 rounds. Furthermore, expended magazines can be quickly replaced, so that a single person with a single assault weapon can easily fire literally hundreds of rounds within minutes. . . . In contrast, hunting rifles and shotguns typically have much smaller magazine capabilities--from 3-5.[64]

Congress specifically exempted 661 long guns from the assault weapon ban that are "most commonly used in hunting and recreational sports."[65]   The vast majority of these long guns do not use large capacity magazines. Although a small number of the exempted long guns have the ability to accept large capacity magazines, only four of these exempted long guns were designed to accept large capacity military magazines.[66]

The 1994 law also demonstrates Congress' concern about the role large capacity magazines and firearms with the ability to accept these large capacity magazines play in

---

[63]   H. Rep. No. 103-489, at 18.

[64]   H. Rep. No. 103-489, at 19 (footnote omitted). The fact that 12 States place a limit on the magazine capacity allowed for hunting, usually 5 or 6 rounds, is consistent with this analysis. (See exhibit 7).

[65]   H. Rep. 103-489, at 20.

[66]   These four firearms are the Iver Johnson M-1 carbine, the Iver Johnson 50th Anniversary M-1 carbine, the Ruger Mini-14 autoloading rifle (without folding stock), and the Ruger Mini Thirty rifle. All of these weapons are manufactured in the United States and are not the subject of this study. In this regard, it should also be noted that Congress can distinguish between domestic firearms and foreign firearms and impose different requirements on the importation of firearms. For example, Congress may ban the importation of certain firearms although similar firearms may be produced domestically. See, for example, B-West Imports v. United States, 75 F.3d 633 (Fed. Cir. 1996).

crime. The House Report for the bill makes reference to numerous crimes involving these magazines and weapons, including the following:[67]

> The 1989 Stockton, California, schoolyard shooting in which a gunman with a semiautomatic copy of an AK47 and 75-round magazines fired 106 rounds in less than 2 minutes. Five children were killed and twenty-nine adults and children were injured.

> The 1993 shooting in a San Francisco, California, office building in which a gunman using 2 TEC DC9 assault pistols with 50-round magazines killed 8 people and wounded 6 others.

> A 1993 shooting on the Long Island Railroad that killed 6 people and wounded 19 others. The gunman had a Ruger semiautomatic pistol, which he reloaded several times with 15-round magazines, firing between 30 to 50 rounds before he was overpowered.

The House Report also includes testimony from a representative of a national police officers' organization, which reflects the congressional concern with criminals' access to firearms that can quickly expel large amounts of ammunition:

> In the past, we used to face criminals armed with a cheap Saturday Night Special that could fire off six rounds before [re]loading. Now it is not at all unusual for a cop to look down the barrel of a TEC-9 with a 32 round clip. The ready availability of and easy access to assault weapons by criminals has increased so dramatically that police forces across the country are being required to upgrade their service weapons merely as a matter of self-defense and preservation. The six-shot .38 caliber service revolver, standard law enforcement issue for years, is just no match against a criminal armed with a semiautomatic assault weapon.[68]

Accordingly, by passing the 1994 law, Congress signaled that firearms with the ability to accept detachable large capacity magazines are not particularly suitable for sporting purposes. Although in 1989 we found the ability to accept a detachable large capacity military magazine was a military configuration feature, we must give it more weight, given this clear signal from Congress.

The passage of the 1994 ban on large capacity magazines has had another effect. Under the 1994 ban, it generally is unlawful to transfer or possess a large capacity magazine

---

[67] H. Rep. No. 103-489, at 15 (two of these examples involve handguns).

[68] H. Rep. 103-489, at 13-14 (footnote omitted).

manufactured after September 13, 1994. Therefore, if we require the LCMM rifles to be modified so that they do not accept a large capacity military magazine in order to be importable, a person will not be able to acquire a newly manufactured large capacity magazine to fit the modified rifle. Thus, the modified rifle neither will be able to accept a grandfathered large capacity military magazine, nor can a new large capacity magazine be manufactured to fit it. Accordingly, today, making the ability to accept a large capacity military magazine disqualifying for importation will prevent the importation of firearms which have the ability to expel large amounts of ammunition quickly without reloading.

This was not the case in 1989 or prior to the 1994 ban.

It is important to note that even though Congress reduced the supply of large capacity military magazines by passing the 1994 ban, there are still vast numbers of grandfathered large capacity military magazines available that can be legally possessed and transferred. These magazines currently fit in the LCMM rifles. Therefore, the 1994 law did not eliminate the need to take further measures to prevent firearms imported into the United States from having the ability to accept large capacity military magazines, a nonsporting factor.

Another impetus for reevaluating the existing standard is the development of modified weapons. The 1989 report caused 43 different models of semiautomatic assault rifles to be banned from being imported into the United States. The effect of that determination was that nearly all semiautomatic rifles with the ability to accept detachable large capacity military magazines were denied importation. Accordingly, at the time, there was no need for the ability to accept such a magazine to be a determining factor in the sporting purposes test. This is no longer the case. As discussed earlier, manufacturers have modified the semiautomatic assault rifles disallowed from importation in 1989 by removing all of their military configuration features, except for the ability to accept a detachable magazine. As a result, semiautomatic rifles with the ability to accept detachable large capacity military magazines (and therefore quickly expel large amounts of ammunition) legally have been entering the United States in significant numbers. Accordingly, the development of these modified weapons necessitates reevaluating our existing standards.

Thus, in order to address Congress' concern with firearms that have the ability to expel large amounts of ammunition quickly, particularly in light of the resumption of these weapons coming into the United States, the ability to accept a detachable large capacity military magazine must be given greater weight in the sporting purposes analysis of the LCMM rifles than it presently receives.[69]

---

[69] A firearm that can be easily modified to accept a detachable large capacity military magazine with only minor adjustments to the firearm or the magazine is considered to be a firearm with the ability to accept these magazines. The ROMAK4 is an example of such a firearm: With minor modifications to either the

Derived from semiautomatic assault rifles that failed to meet the sporting purposes test in 1989 but were later found importable when certain military features were removed.

All rifles that failed to meet the sporting purposes test in 1989 were found to represent a distinctive type of rifle distinguished by certain general characteristics that are common to the modern military assault rifle. Although the LCMM rifles are based on rifle designs excluded from importation under the 1989 standard, they all were approved for import when certain military features were removed. However, the LCMM rifles all still maintain some characteristics common to the modern military assault rifle. Because the outward appearance of most of the LCMM rifles continues to resemble the military assault rifles from which they are derived, we have examined the issue of outward appearance carefully. Some might prefer the rugged, utilitarian look of these rifles to more traditional sporting guns. Others might recoil from using these rifles for sport because of their nontraditional appearance. In the end, we concluded that appearance alone does not affect the LCMM rifles' suitability for sporting purposes. Available information leads us to believe that the determining factor for their use in crime is the ability to accept a detachable large capacity military magazine.

<u>Use</u>

In the 1989 study, ATF found that all rifles fairly typed as semiautomatic assault rifles should be treated the same. Accordingly, the report stated "[t]he fact that there may be some evidence that a particular rifle of this type is used or recommended for sporting purposes should not control its importability. Rather, all findings as to suitability of these rifles as a whole should govern each rifle within this type."[70] We adopt the same approach for the present study.

Use for hunting:

The information we collected on the actual use of the LCMM rifles for hunting medium or larger game suggests that, with certain exceptions, the LCMM rifles sometimes are used for hunting; however, their actual use in hunting is limited.[71] In fact, there are some

---

firearm or a large capacity magazine that was originally designed and produced for a semiautomatic assault rifle based on the AK47 design, the ROMAK4 has the ability to accept the magazine.

[70]   1989 report at 11.

[71]   We targeted the surveys toward the hunting of medium and larger game (e.g., turkey and deer) because the LCMM rifles chamber centerfire cartridges and therefore likely would be most suitable for hunting this type of game. We also learned that the LCMM rifles were used to shoot certain varmints (e.g., coyotes and groundhogs), which are generally considered to be pests, not game. Many commented that the LCMM

general restrictions and prohibitions on the use of any semiautomatic rifle for hunting game. Almost half of the States place restrictions on the use of semiautomatic rifles in hunting, mostly involving magazine capacity (5-6 rounds) and what can be hunted with the rifles (see exhibit 7).

Of the 198 hunting guides who responded to our survey, only 26 stated that they had clients who used the LCMM rifles on hunting trips during the past 2 hunting seasons and only 10 indicated that they recommend the LCMM rifles for hunting. In contrast, the vast majority of the guides (152) indicated that none of their clients used the LCMM rifles on hunting trips during the past 2 hunting seasons. In addition, the hunting guides indicated that the most common semiautomatic rifles used by their clients were those made by Browning and Remington.[72] We found significant the comments of the hunting guides indicating that the LCMM rifles were not widely used for hunting.

Of the 13 editors of hunting and shooting magazines who responded to our survey, only 2 stated that their publications recommend specific types of centerfire semiautomatic rifles for use in hunting medium or larger game. These two respondents stated that they recommend all rifles that are safe and of appropriate caliber for hunting, including the LCMM rifles. However, they did not recommend the LCMM rifles based on the Uzi design for hunting big game; these rifles use a 9mm cartridge, which is not an appropriate caliber for this type of game, according to the editors. It is important to note that the LCMM rifles use different cartridges. The LCMM rifles based on the FN-FAL, SIG SG550, and HK91 and 93 designs are chambered for either the .308 Winchester cartridge or the .223 Remington cartridge, depending on the specific model; the LCMM rifles based on the Uzi design are chambered for the 9mm Parabellum cartridge; and the majority of the LCMM rifles based on the AK47 design are chambered for the 7.62 x 39mm cartridge (some are chambered for the .223 Remington cartridge).

Of the five interest and information groups that responded to our survey, three supported the use of the LCMM rifles for hunting. However, one of these groups stated that the

---

rifles were particularly useful on farms and ranches because of their ruggedness, utilitarian design, and reliability.

[72] According to a 1996 study conducted for the Fish and Wildlife Service, only 2 percent of big game hunters surveyed used licensed hunting guides. Therefore, it should be noted that the information provided by the guides we surveyed may not be representative of all hunters. However, we believe that the hunting guides' information is reliable and instructive because of their high degree of experience with and knowledge of hunting.

ammunition used by the LCMM rifle models based on the Uzi design were inadequate for shooting at long distances (i.e., more than 100 yards).

Out of the 70 published articles reviewed from various shooting magazines, only 5 contained relevant information. One of these five articles stated that, in the appropriate calibers, the LCMM rifles could make "excellent" hunting rifles. Two of the articles stated that the 7.62 x 39mm cartridge (used in LCMM rifles based on the AK47 design) could be an effective hunting cartridge. One of the articles that recommended the rifles also recommended modifications needed to improve their performance in hunting. None of the articles suggested that LCMM rifles based on the Uzi design were good hunting rifles. Thus, although the LCMM rifles could be used in hunting, the articles provided limited recommendations for their use as hunting weapons.

In their usage guides, ammunition manufacturers recommend the .308 and the 7.62 x 39mm cartridges (used in LCMM rifles based on the FN-FAL and HK 91 designs, and the AK47 design respectively) for medium game hunting. However, the usage guides do not identify the 9mm cartridge (used in the Uzi design rifles) as being suitable for hunting.

A majority of the importers who provided information said that the LCMM rifles they import are used for hunting deer and similar animals. However, they provided little evidence that the rifles were especially suitable for hunting these animals. Two of the importers who responded also provided input from citizens in the form of letters supporting this position. The letters show a wide variety of uses for the LCMM rifles, including deer hunting, plinking, target shooting, home defense, and competitive shooting.

Our review of all of this information indicates that while these rifles are used for hunting medium and larger game, as well as for shooting varmints, the evidence was not persuasive that there was widespread use for hunting. We did not find any evidence that the ability to accept a large capacity military magazine serves any hunting purpose. Traditional hunting rifles have much smaller magazine capabilities. Furthermore, the mere fact that the LCMM rifles are used for hunting does not mean that they are particularly suitable for hunting or meet the test for importation.

Use for organized competitive target shooting:

Of the 31 competitive shooting groups we surveyed that stated they have events using high-power semiautomatic rifles, 18 groups stated that they permit the use of the LCMM rifles for all competitions. However, 13 respondents stated that they restrict or prohibit the LCMM rifles for some competitions, and one group stated that it prohibits the LCMM rifles for all competitions. These restrictions and prohibitions generally were enacted for the following reasons:

1. High-power rifle competitions generally require accuracy at ranges beyond the capabilities of the 9mm cartridge, which is used by the LCMM rifles based on the Uzi design.

2. The models based on the AK47 design are limited to competitions of 200 yards or less because the 7.62 x 39mm cartridge, which is used by these models, generally has an effective range only between 300 and 500 yards.

3. Certain matches require U.S. military service rifles, and none of the LCMM rifles fall into this category.

The LCMM rifles are permitted in all United States Practical Shooting Association (USPSA) rifle competitions. The USPSA Practical Shooting Handbook, Glossary of Terms, states that "[y]ou can use any safe firearm meeting the minimum caliber (9mm/.38) and power factor (125PF) requirements." The USPSA has stated that "rifles with designs based on the AR15, AK47, FN-FAL, HK91, HK93, and others are allowed and must be used to be competitive." Moreover, we received some information indicating that the LCMM rifles actually are used in practical shooting competitions.[73] However, we did not receive any information demonstrating that an LCMM rifle's ability to accept large capacity military magazines was necessary for its use in practical shooting competitions.

A couple of the interest groups recommended the LCMM rifles for organized competitive target shooting.

None of the 70 published articles read mentioned the use of the LCMM rifles in organized competitive target shooting.

All of the major ammunition manufacturers produce .308 Winchester ammunition (which is used in the LCMM rifle models based on the HK 91 and FN-FAL designs) and .223 Remington ammunition (which is used in the HK 93, the SIG SG550, and some of the study rifle models based on the AK47 design) specifically for competitive shooting for rifles. The major manufacturers and advertisers of 9mm ammunition (which is used in the LCMM rifles based on the Uzi design) identify it as being suitable for pistol target shooting and self-defense.

---

[73] Merely because a rifle is used in a sporting competition, the rifle does not become a sporting rifle. 114 Cong. Rec. 27461-462 (1968).

A majority of the importers who provided information stated that the LCMM rifles they import are permitted in and suitable for organized competitive target shooting. Two of the importers who responded also provided input from citizens in the form of letters and petitions supporting this position. However, the importers provided little evidence that the rifles were especially suitable for organized competitive target shooting.

The information collected on the actual use of the LCMM rifles for organized competitive target shooting suggests that, with certain exceptions, the LCMM rifles usually may be used and sometimes are used for organized competitive target shooting; however, their suitability for this activity is limited. In fact, there are some restrictions and prohibitions on their use. The use of the rifles in competitive target shooting appears more widespread than for hunting and their use for practical shooting was the most significant. Although we are not convinced that practical shooting does in fact constitute a sporting purpose under section 925(d), we note that there was no information demonstrating that rifles with the ability to accept detachable large capacity military magazines were necessary for use in practical shooting. Once again, the presence of this military feature on LCMM rifles suggests that they are not generally recognized as particularly suitable for or readily adaptable to sporting purposes.

Use in crime:

To fully understand how the LCMM rifles are used, we also examined information available to us on their use in crime. Some disturbing trends can be identified, and it is clear the LCMM rifles are attractive to criminals.

The use of LCMM rifles in violent crime and firearms trafficking is reflected in the cases cited below. It should be noted that the vast majority of LCMM rifles imported during the period 1991-1997 were AK47 variants, which explains their prevalence in the cited cases.

North Philadelphia, Pennsylvania

From April 1995 to November 1996, a convicted felon used a straw purchaser to acquire at least 55 rifles, including a number of MAK90s. The rifles were then trafficked by the prohibited subject to individuals in areas known for their high crime rates. In one case, the rifles were sold from the parking lot of a local elementary school.
Oakland, California

On July 8, 1995, a 32-year-old Oakland police officer assisted a fellow officer with a vehicle stop in a residential area. As the first officer searched the rear compartment of the stopped vehicle, a subject from a nearby residence used a Norinco model NMH 90 to shoot the 32-year old officer in the back. The officer later died from the wound.

El Paso, Texas

On April 15, 1996, after receiving information from the National Tracing Center, ATF initiated an undercover investigation of a suspected firearms trafficker who had purchased 326 MAK90 semiautomatic rifles during a 6-month period. The individual was found to be responsible for illegally diverting more than 1,000 firearms over the past several years. One of the MAK90 rifles that the subject had purchased was recovered from the scene of a 1996 shootout in Guadalajara, Mexico, between suspected drug traffickers and Mexican authorities. Another MAK90 was recovered in 1997 from the residence of a former Mexican drug kingpin following his arrest for drug-related activities.

Charlotte, North Carolina

On May 24, 1996, four armed subjects—one with a MAK90 rifle—carried out a home invasion robbery during which they killed the resident with a 9mm pistol. All four suspects were arrested.

Dallas, Texas

In September 1997, an investigation was initiated on individuals distributing crack cocaine from a federally subsidized housing community. During repeated undercover purchases of the narcotics, law enforcement officials noticed that the suspects had firearms in their possession. A search warrant resulted in the seizure of crack cocaine, a shotgun, and a North China Industries model 320 rifle.

Chesterfield, Virginia

In November 1997, a MAK90 rifle was used to kill two individuals and wound three others at a party in Chesterfield, Virginia.

Orange, California

In December 1997, a man armed with an AKS 762 rifle and two other guns drove to where he was previously employed and opened fire on former coworkers, killing four and injuring three, including a police officer.

Baltimore, Maryland

In December 1997, a search warrant was served on a homicide suspect who was armed at the time with three pistols and a MAK90 rifle.

We also studied import and trace information to learn whether the LCMM rifles are used in crime.

Between 1991 and 1997, there were 425,114 LCMM rifles imported into the United States. This represents 7.6 percent of the approximately 5 million rifles imported during this period. The breakdown of the specific variants of LCMM rifles imported follows:

AK-47 variants: 377,934
FN-FAL variants: 37,534
HK variants: 6,495
Uzi variants: 3,141
SIG SG550 variants: 10

During this same time period, ATF traced 632,802 firearms.[74] This included 81,842 rifles of which approximately 3,176 were LCMM rifles.[75] While this number is relatively low compared to the number of total traces, it must be viewed in light of the small number of LCMM rifles imported during this time period and the total number of rifles, both imported domestic, that were available in the United States. A more significant trend is reflected in figure 1.

---

[74] ATF traces crime guns recovered and submitted by law enforcement officials. A crime gun is defined, for purposes of firearms tracing, as any firearm that is illegally possessed, used in a crime, or suspected by law enforcement of being used in a crime. Trace information is used to establish links between criminals and firearms, to investigate illegal firearm trafficking, and to identify patterns of crime gun traces by jurisdiction. A substantial number of firearms used in crime are not recovered by law enforcement agencies and therefore not traced. In addition, not all recovered crime guns are traced. Therefore, trace requests substantially underestimate the number of firearms involved in crimes, and trace numbers contain unknown statistical biases. These problems are being reduced as more law enforcement agencies institute policies of comprehensive crime gun tracing.

[75] The vast majority of LCMM rifles traced during this time period were AK47 variants. Specifically, AK47 variants comprised 95.6 percent of the LCMM rifles traced. This must be viewed within the context that 88 percent of the LCMM rifles imported during this period were AK47 variants.

Firearms Traces 1991-1997

Year	Total Firearms Traced	Total Rifles Traced	Total Assault[76] Rifles Traced	Total LCMM Rifles Traced
1991	42,442	6,196	656	7
1992	45,134	6,659	663	39
1993	54,945	7,690	852	182
1994	83,137	9,201	735	596
1995	76,847	9,988	717	528
1996	136,062	17,475	1,075	800
1997	194,235	24,633	1,518	1,024
Cumulative Total	632,802	81,842	6,216	3,176

Figure 1

The figures in this table show that between 1991 and 1994, trace requests involving LCMM rifles increased rapidly, from 7 to 596. During the same period, trace requests for assault rifles increased at a slower rate, from 656 to 735. The years 1991 to 1994 are significant because they cover a period between when the ban on the importation of semiautomatic assault rifles was imposed and before the September 13, 1994, ban on semiautomatic assault weapons was enacted. Thus, during the years leading up to the 1994 ban, traces of LCMM rifles were increasing much more rapidly than the traces of the rifles that had been the focus of the 1989 ban, as well as the rifles that were the focus of the 1994 congressional action.

We also compared patterns of importation with trace requests to assess the association of LCMM rifles with criminal involvement. The comparison shows that importation of LCMM rifles in the early 1990s was followed immediately by a rapid rise in the number of trace requests involving LCMM rifles. This is shown in figures 2 and 3.

[76] For purposes of this table, assault rifles include (1) semiautomatic assault rifles banned from importation in 1989 but still available domestically because they had been imported into the United States prior to the ban, (2) domestically produced rifles that would not have qualified for importation after 1989, and (3) semiautomatic assault rifles that were banned in 1994.

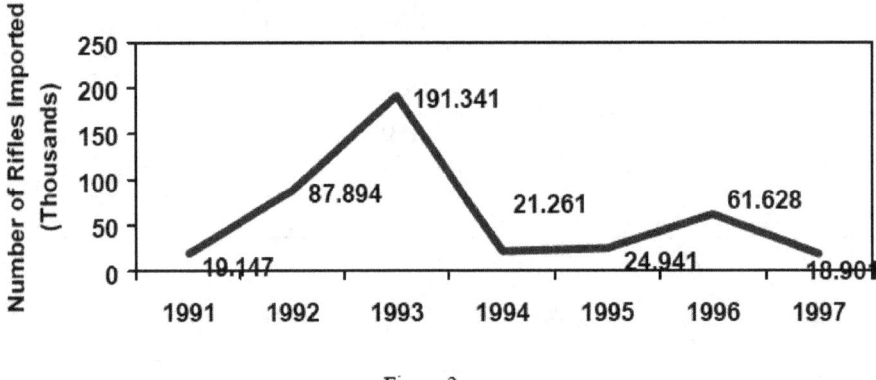

Figure 2

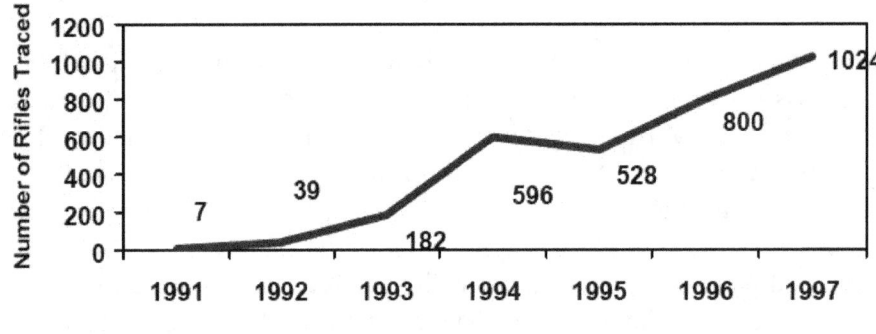

Figure 3

Two aspects of the relationship between importation and trace request patterns are significant. First, the rapid rise in traces following importation indicates that, at least in some cases, very little time elapsed between a particular LCMM rifle's importation and its recovery by law enforcement. This time lapse is known as "time to crime." A short time to crime can be an indicator of illegal trafficking. Therefore, trace patterns suggest what the case examples show: LCMM rifles have been associated with illegal trafficking. Second, while LCMM rifles have not been imported in large numbers since 1994,[77] the number of trace requests for LCMM rifles continues to rise. This reflects a sustained and

---

[77]   One reason is that there has been an embargo on the importation of firearms from China since May 1994.

continuing pattern of criminal association for LCMM rifles despite the fact that there were fewer new LCMM rifles available.[78] Moreover, it is reasonable to conclude that if the importation of LCMM rifles resumes, the new rifles would contribute to the continuing rise in trace requests for them.[79]

All of the LCMM rifles have the ability to accept a detachable large capacity military magazine. Thus, they all have the ability to expend large amounts of ammunition quickly. In passing the 1994 ban on semiautomatic assault rifles and large capacity ammunition feeding devices, Congress found that weapons with this ability are attractive to criminals.[80] Thus, we can infer that the LCMM rifles may be attractive to criminals because in some ways they remain akin to military assault rifles, particularly in their ability to accept a detachable large capacity military magazine.

---

[78] The increase in trace requests also reflects the fact that law enforcement officials were making trace requests for all types of firearms much more frequently beginning in 1996. There were 76,847 trace requests in 1995, 136,062 trace requests in 1996, and 194,235 trace requests in 1997. Traces for assault rifles were increasing by approximately the same percentage as traces for LCMM rifles during these years.

[79] In addition to looking at case studies and tracing and import information, we attempted to get information on the use of the LCMM rifles in crime by surveying national law enforcement agencies and organizations, as well as metropolitan police departments. Twenty-three national law enforcement agencies and organizations were surveyed and five responded. Three of the respondents stated they had no information. The other two provided information that was either outdated or not specific enough to identify the LCMM rifles

The 26 metropolitan police departments surveyed provided the following information:

17 departments had no information to provide.
5 departments stated that the LCMM rifles were viewed as crime guns.
1 department stated that the LCMM rifles were nonsporting.
2 departments stated that the LCMM rifles were used to hunt coyotes in their areas.
1 department stated that the LCMM rifles were used for silhouette target shooting.

[80] H. Rep. No. 103-489, at 13, 18, 19.

DETERMINATION

In 1989, ATF determined that the type of rifle defined as a semiautomatic assault rifle was not generally recognized as particularly suitable for or readily adaptable to sporting purposes. Accordingly, ATF found that semiautomatic assault rifles were not importable into the United States. This finding was based, in large part, on ATF's determination that semiautomatic assault rifles contain certain general characteristics that are common to the modern military assault rifle. These characteristics were designed for killing and disabling the enemy and distinguish the rifles from traditional sporting rifles. One of these characteristics is a military configuration, which incorporates eight physical features: Ability to accept a detachable magazine, folding/telescoping stocks, separate pistol grips, ability to accept a bayonet, flash suppressors, bipods, grenade launchers, and night sights. In 1989, ATF decided that any of these military configuration features, other than the ability to accept a detachable magazine, would make a semiautomatic assault rifle not importable.

Certain semiautomatic assault rifles that failed the 1989 sporting purposes test were modified to remove all of the military configuration features, except for the ability to accept a detachable magazine. Significantly, most of these modified rifles not only still have the ability to accept a detachable magazine but, more specifically, still have the ability to accept a large capacity military magazine. It appears that only one of the current study rifles, the VEPR caliber .308 (an AK47 variant), does not have the ability to accept a large capacity military magazine and, therefore, is not an LCMM rifle. Based on the standard developed in 1989, these modified rifles were found not to fall within the semiautomatic assault rifle type and were found to meet the sporting purposes test. Accordingly, these rifles were approved for import into the United States.

Members of Congress and others have expressed concerns that these modified semiautomatic assault rifles are essentially the same as the semiautomatic assault rifles determined to be not importable in 1989. In response to such concerns, the present study reviewed the current application of the sporting purposes test to the study rifles to determine whether the statute is being applied correctly and to ensure that the current use of the study rifles is consistent with the statute's criteria for importability.

Our review took another look at the entire matter. We reexamined the basic tenets of the 1989 study, conducted a new analysis of the physical features of the rifles, surveyed a wide variety of sources to acquire updated information relating to use and suitability, and assessed changes in law that might have bearing on the treatment of the study rifles.

This review has led us to conclude that the basic finding of the 1989 decision remains valid and that military-style semiautomatic rifles are not importable under the sporting purposes standard. Accordingly, we believe that the Department of the Treasury correctly has been denying the importation of rifles that had any of the distinctly military configuration features identified in 1989, other than the ability to accept a detachable magazine. Our review, however, did result in a finding that the ability to accept a detachable large capacity magazine originally designed and produced for a military assault weapon should be added to the list of disqualifying military configuration features identified in 1989.

Several important changes have occurred since 1989 that have led us to reevaluate the importance of this feature in the sporting purposes test. Most significantly, by passing the 1994 bans on semiautomatic assault weapons and large capacity ammunition feeding devices, Congress sent a strong signal that firearms with the ability to expel large amounts of ammunition quickly are not sporting; rather, firearms with this ability have military purposes and are a crime problem. The House Report to the 1994 law emphasizes that the ability to accept a large capacity magazine "serve[s] specific, combat-functional ends."[81] Moreover, this ability plays a role in increasing a firearm's "capability for lethality," creating "more wounds, more serious, in more victims."[82] Furthermore, the House Report noted semiautomatic assault weapons with this ability are the "weapons of choice among drug dealers, criminal gangs, hate groups, and mentally deranged persons bent on mass murder."[83]

Moreover, we did not find any evidence that the ability to accept a detachable large capacity military magazine serves any sporting purpose. The House Report to the 1994 law notes that, while most of the weapons covered by the assault weapon ban come equipped with detachable large capacity magazines, hunting rifles and shotguns typically have much smaller magazine capabilities, from 3 to 5 rounds.[84] Similarly, we found that a number of States limit magazine capacity for hunting to 5 to 6 rounds. We simply found no information showing that the ability to accept a detachable large capacity military magazine has any purpose in hunting or organized competitive target shooting.

Accordingly, we find that the ability to accept a detachable large capacity military magazine is a critical factor in the sporting purposes test that must be given the same weight as the other military configuration features identified in 1989.

The information we collected on the use and suitability of the LCMM rifles for hunting and organized competitive target shooting demonstrated that the rifles are not especially suitable for sporting purposes. Although our study found that the LCMM rifles, as a type, may sometimes be used for hunting, we found no evidence that they are commonly used for hunting. In fact, some of the rifles are unsuitable for certain types of hunting.

---

[81] H. Rep. No. 103-489, at 18.

[82] H. Rep. No. 103-489, at 19.

[83] H. Rep. No. 103-489, at 13.

[84] H. Rep. No. 103-489, at 19 (footnote omitted).

The information we collected also demonstrated that although the LCMM rifles, as a type, may be used for organized competitive target shooting, their suitability for these competitions is limited. There are even some restrictions or prohibitions on their use for certain types of competitions. In addition, we believe that all rifles which are fairly typed as LCMM rifles should be treated the same. Therefore, the fact that there may be some evidence that a particular rifle of this type is used or recommended for sporting purposes should not control its importability. Rather, all findings as to suitability of LCMM rifles as a whole should govern each rifle within this type. The findings as a whole simply did not satisfy the standard set forth in section 925(d)(3).

Finally, the information we gathered demonstrates that the LCMM rifles are attractive to certain criminals. We find that the LCMM rifles' ability to accept a detachable large capacity military magazine likely plays a role in their appeal to these criminals. In enacting the 1994 bans on semiautomatic assault weapons and large capacity ammunition feeding devices, Congress recognized the appeal large magazine capacity has to the criminal element.

Weighing all this information, the LCMM rifles, as a type, are not generally recognized as particularly suitable for or readily adaptable to sporting purposes. As ATF found in conducting its 1989 study, although some of the issues we confronted were difficult to resolve, in the end we believe the ultimate conclusion is clear and compelling. The ability of all of the LCMM rifles to accept a detachable large capacity military magazine gives them the capability to expel large amounts of ammunition quickly; this serves a function in combat and crime, but serves no sporting purpose. Given the high standard set forth in section 925(d)(3) and the Secretary's discretion in applying the sporting purposes test, this conclusion was clear.

This decision will in no way preclude the importation of true sporting firearms. It will prevent only the importation of firearms that cannot fairly be characterized as sporting rifles.

Individual importers with existing permits for, and applications to import involving, the LCMM rifles will be notified of this determination in writing. Each of these importers will be given an opportunity to respond and present additional information and arguments. Final action will be taken on permits and applications only after an affected importer has an opportunity to makes its case.

THE WHITE HOUSE
WASHINGTON

November 14, 1997

MEMORANDUM FOR THE SECRETARY OF THE TREASURY

SUBJECT: Importation of Modified Semiautomatic
                Assault-Type Rifles

The Gun Control Act of 1968 restricts the importation of
firearms unless they are determined to be particularly suitable
for or readily adaptable to sporting purposes. In 1989, the
Department of the Treasury (the Department) conducted a review
of existing criteria for applying the statutory test based on
changing patterns of gun use. As a result of that review,
43 assault-type rifles were specifically banned from importation. However, manufacturers have modified many of those
weapons banned in 1989 to remove certain military features
without changing their essential operational mechanism.
Examples of such weapons are the Galil and the Uzi.

In recent weeks Members of Congress have strongly urged that it
is again necessary to review the manner in which the Department
is applying the sporting purposes test, in order to ensure that
the agency's practice is consistent with the statute and current
patterns of gun use. A letter signed by 30 Senators strongly
urged that modified assault-type weapons are not properly
importable under the statute and that I should use my authority
to suspend temporarily their importation while the Department
conducts an intensive, expedited review. A recent letter from
Senator Dianne Feinstein emphasized again that weapons of this
type are designed not for sporting purposes but for the commission of crime. In addition, 34 Members of the House of
Representatives signed a letter to Israeli Prime Minister
Binyamin Netanyahu requesting that he intervene to stop all
sales of Galils and Uzis into the United States. These
concerns have caused the Government of Israel to announce
a temporary moratorium on the exportation of Galils and Uzis
so that the United States can review the importability of
these weapons under the Gun Control Act.

Exhibit 1

2

The number of weapons at issue underscores the potential threat to the public health and safety that necessitates immediate action. Firearms importers have obtained permits to import nearly 600,000 modified assault-type rifles. In addition, there are pending before the Department applications to import more than 1 million additional such weapons. The number of rifles covered by outstanding permits is comparable to that which existed in 1989 when the Bush Administration temporarily suspended import permits for assault-type rifles. The number of weapons for which permits for importation are being sought through pending applications is approximately 10 times greater than in 1989. The number of such firearms for which import applications have been filed has skyrocketed from 10,000 on October 9, 1997, to more than 1 million today.

My Administration is committed to enforcing the statutory restrictions on importation of firearms that do not meet the sporting purposes test. It is necessary that we ensure that the statute is being correctly applied and that the current use of these modified weapons is consistent with the statute's criteria for importability. This review should be conducted at once on an expedited basis. The review is directed to weapons such as the Uzi and Galil that failed to meet the sporting purposes test in 1989, but were later found importable when certain military features were removed. The results of this review should be applied to all pending and future applications.

The existence of outstanding permits for nearly 600,000 modified assault-type rifles threatens to defeat the purpose of the expedited review unless, as in 1989, the Department temporarily suspends such permits. Importers typically obtain authorization to import firearms in far greater numbers than are actually imported into the United States. However, gun importers could effectively negate the impact of any Department determination by simply importing weapons to the maximum amount allowed by their permits. The public health and safety require that the only firearms allowed into the United States are those that meet the criteria of the statute.

Accordingly, as we discussed, you will:

    1) Conduct an immediate expedited review not to exceed 120 days in length to determine whether modified semiautomatic assault-type rifles are properly importable under the statutory sporting purposes test. The results of this review will govern action on pending and future applications for import permits, which shall not be acted upon until the completion of this review.

Exhibit 1

3

2) Suspend outstanding permits for importation of modified semiautomatic assault-type rifles for the duration of the 120-day review period. The temporary suspension does not constitute a permanent revocation of any license. Permits will be revoked only if and to the extent that you determine that a particular weapon does not satisfy the statutory test for importation, and only after an affected importer has an opportunity to make its case to the Department.

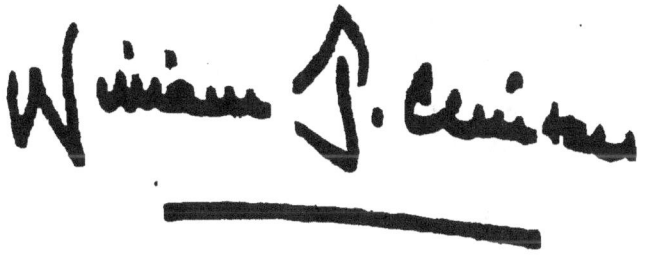

Exhibit 2

STUDY RIFLE MODELS

AK47 Variants: FN-FAL Variants:

MAK90*	SA2000	Saiga rifle	L1A1 Sporter
314*	ARM	Galil Sporter	FAL Sporter
56V*	MISR	Haddar	FZSA
89*	MISTR	Haddar II	SAR4800
EXP56A*	SA85M	WUM 1	X FAL
SLG74	Mini PSL	WUM 2	C3
NHM90*	ROMAK 1	SLR95	C3A
NHM90-2*	ROMAK 2	SLR96	LAR Sporter
NHM91*	ROMAK 4	SLR97	
SA85M	Hunter rifle	SLG94	
SA93	386S	SLG95	
A93	PS/K	SLG96	
AKS 762	VEPR caliber		
VEPR caliber .308	7.62 x 39mm		

HK Variants: Uzi Variants: SIG SG550 Variants:

BT96	Officers 9*	SG550-1
Centurian 2000	320 carbine*	SG550-2
SR9	Uzi Sporter	
PSG1		
MSG90		
G3SA		
SAR8		

- These models were manufactured in China and have not been imported since the 1994 embargo on the importation of firearms from China.

Exhibit 3

STUDY RIFLES

The study rifles are semiautomatic firearms based on the AK47, FN-FAL, HK 91 and 93, Uzi, and SIG SG550 designs. Each of the study rifles is derived from a semiautomatic assault rifle. The following are some examples of specific study rifle models grouped by design type. In each instance, a semiautomatic assault rifle is shown above the study rifles for comparison.

AK47 Variants

AK47 semiautomatic assault rifle

===

MISR   ARM

MAK90   WUM 1

Exhibit 3

FN-FAL Variants

FN-FAL semiautomatic assault rifle

===

L1A1 Sporter   SAR 4800

HK 91 and 93 Variants

HK91 semiautomatic assault rifle

===

SR9   SAR 8

Exhibit 3

Uzi Variants

Uzi semiautomatic assault rifle

===

320 carbine

SIG SG550 Variants

The following illustration depicts the configuration of a semiautomatic assault rifle based on the SIG SG550 design. No illustrations of modified semiautomatic versions are available.

SIG SG550 semiautomatic assault rifle

Exhibit 4

DEPARTMENT OF THE TREASURY
BUREAU OF ALCOHOL, TOBACCO AND FIREARMS

FACTORING CRITERIA FOR WEAPONS

NOTE: The Bureau of Alcohol, Tobacco and Firearms reserves the right to preclude importation of any revolver or pistol which achieves an apparent qualifying score but does not adhere to the provisions of section 925(d)(3) of Amended Chapter 44, Title 18, U.S.C.

PISTOL	REVOLVER
MODEL:	MODEL:

PREREQUISITES (PISTOL)

1. The pistol must have a positive manually operated safety device.
2. The combined length and height must not be less than 10" with the height (right angle measurement to barrel without magazine or extension) being at least 4" and the length being at least 6"

PREREQUISITES (REVOLVER)

1. Must pass safety test.
2. Must have overall frame (with conventional grips) length (not diagonal) of 4½" minimum.
3. Must have a barrel length of at least 3".

INDIVIDUAL CHARACTERISTICS	POINT VALUE	POINT SUB-TOTAL	INDIVIDUAL CHARACTERISTICS	POINT VALUE	POINT SUB-TOTAL
OVERALL LENGTH			BARREL LENGTH (Muzzle to Cylinder Face)		
FOR EACH 1/4" OVER 6"	1		LESS THAN 4"	0	
FRAME CONSTRUCTION			FOR EACH 1/4" OVER 4"	1/2	
INVESTMENT CAST OR FORGED STEEL	15		FRAME CONSTRUCTION		
INVESTMENT CAST OR FORGED HTS ALLOY	20		INVESTMENT CAST OR FORGED STEEL	15	
WEAPON WEIGHT W/MAGAZINE (Unloaded)			INVESTMENT CAST OR FORGED HTS ALLOY	20	
PER OUNCE	1		WEAPON WEIGHT (Unloaded)		
CALIBER			PER OUNCE	1	
.22 SHORT AND .25 AUTO	0		CALIBER		
.22 LR AND 7.65mm TO .380 AUTO	3		.22 SHORT TO .25 ACP	0	
9mm PARABELLUM AND OVER	10		.22 LR AND .30 TO .38 S&W	3	
SAFETY FEATURES			.38 SPECIAL	4	
LOCKED BREECH MECHANISM	5		.357 MAG AND OVER	5	
LOADED CHAMBER INDICATOR	5		MISCELLANEOUS EQUIPMENT		
GRIP SAFETY	3		ADJUSTABLE TARGET SIGHTS (Drift or Click)	5	
MAGAZINE SAFETY	5		TARGET GRIPS	5	
FIRING PIN BLOCK OR LOCK	10		TARGET HAMMER AND TARGET TRIGGER	5	
MISCELLANEOUS EQUIPMENT			SAFETY TEST		
EXTERNAL HAMMER	2		A Double Action Revolver must have a safety feature which automatically (or in a Single Action Revolver by manual operation) causes the hammer to retract to a point where the firing pin does not rest upon the primer of the cartridge. The safety device must withstand the impact of a weight equal to the weight of the revolver dropping from a distance of 36" in a line parallel to the barrel upon the rear of the hammer spur, a total of 5 times.		
DOUBLE ACTION	10				
DRIFT ADJUSTABLE TARGET SIGHT	5				
CLICK ADJUSTABLE TARGET SIGHT	10				
TARGET GRIPS	5				
TARGET TRIGGER	2				
SCORE ACHIEVED (Qualifying score is 75 points)			SCORE ACHIEVED (Qualifying score is 45 points)		

Exhibit 5

MILITARY CONFIGURATION

1. <u>Ability to accept a detachable magazine</u>. Virtually all modern military firearms are designed to accept large, detachable magazines. This provides the soldier with a fairly large ammunition supply and the ability to rapidly reload. Thus, large capacity magazines are indicative of military firearms. While detachable magazines are not limited to military firearms, most traditional semiautomatic sporting firearms, designed to accommodate a detachable magazine, have a relatively small magazine capacity. Additionally, some States have a limit on the magazine capacity allowed for hunting, usually five or six rounds.

2. <u>Folding/telescoping stock</u>. Many military firearms incorporate folding or telescoping stocks. The main advantage of this item is portability, especially for airborne troops. These stocks allow the firearm to be fired from the folded position, yet it cannot be fired nearly as accurately as with an open stock. With respect to possible sporting uses of this feature, the folding stock makes it easier to carry the firearm when hiking or backpacking. However, its predominant advantage is for military purposes, and it is normally not found on the traditional sporting rifle.

3. <u>Pistol grips</u>. The vast majority of military firearms employ a well-defined separate pistol grip that protrudes conspicuously beneath the action of the weapon. In most cases, the "straight line design" of the military weapon dictates a grip of this type so that the shooter can hold and fire the weapon. Further, a pistol grip can be an aid in one-handed firing of the weapon in a combat situation. Further, such grips were designed to assist in controlling machineguns during automatic fire. On the other hand, the vast majority of sporting firearms employ a more traditional pistol grip built into the wrist of the stock of the firearm since one-handed shooting is not usually employed in hunting or organized competitive target competitions.

4. <u>Ability to accept a bayonet</u>. A bayonet has distinct military purposes. First, it has a psychological effect on the enemy. Second, it enables soldiers to fight in close quarters with a knife attached to their rifles. No traditional sporting use could be identified for a bayonet.

5. <u>Flash suppressor</u>. A flash suppressor generally serves one or two functions. First, in military firearms it disperses the muzzle flash when the firearm is fired to help conceal the shooter's position, especially at night. A second purpose of some flash suppressors is to assist in controlling the "muzzle climb" of the rifle, particularly when fired as a fully automatic weapon. From the standpoint of a traditional sporting firearm, there is no particular benefit in suppressing muzzle flash. Flash suppressors that also serve to dampen muzzle climb have a limited benefit in sporting uses by allowing the shooter to reacquire

Exhibit 5

the target for a second shot. However, the barrel of a sporting rifle can be modified by "magna-porting" to achieve the same result. There are also muzzle attachments for sporting firearms to assist in the reduction of muzzle climb. In the case of military-style weapons that have flash suppressors incorporated in their design, the mere removal of the flash suppressor may have an adverse impact on the accuracy of the firearm.

6. <u>Bipods</u>. The majority of military firearms have bipods as an integral part of the firearm or contain specific mounting points to which bipods may be attached. The military utility of the bipod is primarily to provide stability and support for the weapon when fired from the prone position, especially when fired as a fully automatic weapon. Bipods are available accessory items for sporting rifles and are used primarily in long-range shooting to enhance stability. However, traditional sporting rifles generally do not come equipped with bipods, nor are they specifically designed to accommodate them. Instead, bipods for sporting firearms are generally designed to attach to a detachable "slingswivel mount" or simply clamp onto the firearm.

7. <u>Grenade launcher</u>. Grenade launchers are incorporated in the majority of military firearms as a device to facilitate the launching of explosive grenades. Such launchers are generally of two types. The first type is a flash suppressor designed to function as a grenade launcher. The second type attaches to the barrel of the rifle by either screws or clamps. No traditional sporting application could be identified for a grenade launcher.

8. <u>Night sights</u>. Many military firearms are equipped with luminous sights to facilitate sight alignment and target acquisition in poor light or darkness. Their uses are generally for military and law enforcement purposes and are not usually found on sporting firearms since it is generally not legal to hunt at night.

Exhibit 6

[This document has been retyped for clarity.]

MEMORANDUM TO FILE

FIREARMS ADVISORY PANEL

The initial meeting of the Firearms Advisory Panel was held in Room 3313, Internal Revenue Building, on December 10, 1968, with all panel members present. Internal Revenue Service personnel in attendance at the meeting were the Director, Alcohol and Tobacco Tax Division, Harold Serr; Chief, Enforcement Branch, Thomas Casey; Chief, Operations Coordination Section, Cecil M. Wolfe, and Firearms Enforcement Officer, Paul Westenberger. Deputy Assistant Commissioner Compliance, Leon Green, visited the meeting several times during the day.

The Director convened the meeting at 10:00 a.m. by welcoming the members and outlining the need for such an advisory body. He then introduced the Commissioner of Internal Revenue, Mr. Sheldon Cohen, to each panel member.

Mr. Cohen spoke to the panel for approximately fifteen minutes. He thanked the members for their willingness to serve on the panel, explained the role of the panel and some of the background which led to the enactment of the Gun Control Act of 1968. Commissioner Cohen explained to the panel members the conflict of interest provisions of regulations pertaining to persons employed by the Federal Government and requested that if any member had any personal interest in any matter that came under discussion or consideration, he should make such interest known and request to be excused during consideration of the matter.

Mr. Seer then explained to the panel the areas in which the Division would seek the advice of the panel and emphasized that the role of the panel would be advisory only, and that it was the responsibility of the Service to make final decisions. He then turned the meeting over to the moderator, Mr. Wolfe.

Mr. Wolfe explained the responsibility of the Service under the import provisions of the Gun Control Act and under the Mutual Security Act. The import provisions were read and discussed.

The panel was asked to assist in defining Asporting purposes≃ as used in the Act. It was generally agreed that firearms designed and intended for hunting and all types of organized competitive target shooting would fall within the sporting purpose category. A discussion was held on the so-called sport of Aplinking≅. It was the consensus that, while many persons

Exhibit 6

participated in the type of activity and much ammunition was expended in such endeavors, it was primarily a pastime and could not be considered a sport for the purposes of importation since any firearm that could expel a projectile could be used for this purpose without having any characteristics generally associated with target guns.

The point system that had been developed by the Division and another point system formula suggested and furnished by the Southern Gun Distributors through Attorney Michael Desalle, was explained and demonstrated to the panel by Paul Westenberger. Each panel member was given copies of the formulas and requested to study them and endeavor to develop a formula he believed would be equitable and could be applied to all firearms sought to be imported.

A model BM59 Beretta, 7.62 mm, NATO Caliber Sporter Version Rifle was presented to the panel and their advice sought as to their suitability for sporting purposes. It was the consensus that these rifles do have a particular use in target shooting and hunting. Accordingly, it was recommended that importation of this rifle together with the SIG-AMT 7.62mm NATO Caliber Sporting Rifle and the Cetme 7.62mm NATO Caliber Sporting Rifle be authorized for importation. Importation, however, should include the restriction that these weapons must not possess combination flash suppressors/grenade adaptors with outside diameters greater than 20mm (.22 mm is the universal grade adaptor size).

The subject of ammunition was next discussed. Panel members agreed that incendiary and tracer small arms ammunition have no use for sporting purposes. Accordingly, the Internal Revenue Service will not authorize these types of small arms ammunition importation. All other conventional small arms ammunition for pistols, revolvers, rifles and shotguns will be authorized.

The meeting was adjourned at 4:00 p.m.

C.M. Wolfe

Exhibit 7

STATE FISH AND GAME COMMISSION REVIEW

STATE RESTRICTION	RIFLE RESTRICTION	MAGAZINE RESTRICTION
Alabama	Not for turkey	
Alaska		
Arizona		Not more than five rounds
Arkansas	Not for turkey	
California		
Colorado		Not more than six rounds
Connecticut*	No rifles on public land	
Delaware	No rifles	
Florida		Not more than five rounds
Georgia	Not for turkey	
Hawaii		
Idaho	Not for turkey	
Illinois	Not for deer or turkey	
Indiana*	Not for deer or turkey	
Iowa	Not for deer or turkey No restrictions on coyote or fox	
Kansas		
Kentucky		
Louisiana	Not for turkey	
Maine*	Not for turkey	
Maryland*		

Exhibit 7

STATE RESTRICTION	RIFLE RESTRICTION	MAGAZINE RESTRICTION
Massachusetts	Not for deer or turkey	
Michigan	Not for turkey	Not more than six rounds
Minnesota		
Mississippi	Not for turkey	
Missouri	Not for turkey	Chamber and magazine not more than 11 rounds
Montana		
Nebraska		Not more than six rounds
Nevada	Not for turkey	
New Hampshire*	Not for turkey	Not more than five rounds
New Jersey	No rifles	
New Mexico	Not for turkey	
New York*		Not more than six rounds
North Carolina	Not for turkey	
North Dakota	Not for turkey	
Ohio	Not for deer or turkey	
Oklahoma		Not more than seven rounds for .22 caliber
Oregon*		Not more than five rounds
Pennsylvania*	No semiautomatics	

Exhibit 7

STATE RESTRICTION	RIFLE RESTRICTION	MAGAZINE RESTRICTION
Rhode Island	Prohibited except for woodchuck in summer	
South Carolina	Not for turkey	
South Dakota		Not more than five rounds
Tennessee	Not for turkey	
Texas		
Utah	Not for turkey	
Vermont		Not more than six rounds
Virginia*		
Washington	Not for turkey	
West Virginia		
Wisconsin		
Wyoming		

* Limited restrictions (e.g., specified areas, county restrictions, populated areas, time of day).

DEPARTMENT OF THE TREASURY
BUREAU OF ALCOHOL, TOBACCO AND FIREARMS
WASHINGTON, D.C. 20226

DIRECTOR

O:F:S:DMS
3310

Dear Sir or Madam:

On November 14, 1997, the President and the Secretary
of the Treasury decided to conduct a review to
determine whether modified semiautomatic assault rifles
are properly importable under Federal law. Under
18 U.S.C. section 925(d)(3), firearms may be imported
into the United States only if they are determined to
be of a type generally recognized as particularly
suitable for or readily adaptable to sporting purposes.
The firearms in question are semiautomatic rifles based
on the AK47, FN-FAL, HK91, HK93, SIG SG550-1, and Uzi
designs.

As part of the review, the Bureau of Alcohol, Tobacco
and Firearms (ATF) is interested in receiving
information that shows whether any or all of the above
types of semiautomatic rifles are particularly suitable
for or readily adaptable to hunting or organized
competitive target shooting. We are asking that you
voluntarily complete the enclosed survey to assist us
in gathering this information. We anticipate that the
survey will take approximately 15 minutes to complete.

Responses must be received no later than January 9,
1998; those received after that date cannot be included
in the review. Responses should be forwarded to the
Bureau of Alcohol, Tobacco and Firearms, Department HG,
P.O. Box 50860, Washington, DC 20091. We appreciate
any information you care to provide.

Sincerely yours,

John W. Magaw
John W. Magaw
Director

Enclosure

ATF SURVEY OF HUNTING GUIDES
FOR RIFLE USAGE

*Please report only on those clients who **hunted medium game** (for example, turkey) or **larger game** (for example, deer) with a rifle.*

For the purposes of this survey, please count only individual clients and NOT the number of trips taken by a client. For example, if you took the same client on more than one trip, count the client only once.

1. What is the approximate number of your clients who have ever used **manually operated rifles** during the past two hunting seasons of 1995 and 1996?

_____number of clients.

2. What is the approximate number of your clients who have ever used **semiautomatic rifles** during the past two hunting seasons of 1995 and 1996?

_____number of clients.

3. What is the approximate number of your clients who have ever used semiautomatic rifles whose design is based on the **AK 47, FN-FAL, HK91, HK93, SIG 550-1, or Uzi** during the past two hunting seasons of 1995 and 1996?

_____number of clients.

4. From your knowledge, for your clients who use **semiautomatic rifles**, please list the three most commonly used rifles.

<u>Make</u>	<u>Model</u>	<u>Caliber</u>

5. Do you **recommend** the use of any specific rifles by your clients?

_____Yes *(Continue to #6)*   _____No *(You are finished with the survey. Thank you.)*

ATF SURVEY OF HUNTING GUIDES
FOR RIFLE USAGE
Page 2 of 2

6. If your answer to item 5 is "Yes", please identify the specific rifles you **recommend**.

Make	Model	Caliber

7. Do you **recommend** the use of any semiautomatic rifles whose design is based on the **AK 47, FN-FAL, HK91, HK93, SIG 550-1, or Uzi?**

_____Yes *(Continue to #8)*   _____No *(You are finished with the survey. Thank you.)*

8. If your answer to item 7 is "Yes", please identify the specific rifles whose design is based on the **AK 47, FN-FAL, HK91, HK93, SIG 550-1, or Uzi** that you recommend.

Make	Model	Caliber

An agency may not conduct or sponsor, and a person is not required to respond to, the collection of information unless it displays a currently valid OMB control number.

Hunting Guides

case		Number of clients Using			Recommend	
		Manual	Semiauto	AK47 et.al.	Any	AK47 et.al.
A	1	28	0	0	No	
A	2	100	10	0	Yes	No
A	3	18	0	0	No	
A	4	120	40	0	Yes	No
A	5	12	0	0	Yes	No
A	6	80	40	0	No	
A	7	275	25	0	No	
A	8					
A	9	0	0	0		
A	10	0				
A	11	2	5	0	Yes	Yes
A	12	12	0	0	Yes	No
A	13	10	6	0	No	No
A	14	5	7	0	No	
A	15	0	0	0		
A	16	20	0	0	No	No
A	17					
A	18	0	0	0	No	
A	19	17	6	0	No	
A	20	30	8	0	No	
A	21	117	7	0	Yes	No
A	22	160	0	0	Yes	No
A	23	23	1	0	Yes	No
A	24	100	5	0	Yes	No
A	25	210	10	0	Yes	No
A	26	12	4	1	Yes	Yes
A	27	24	3	0	Yes	No
A	28	20	15	0	Yes	No
A	29	4	0	0	No	No
A	30	4	0	0	Yes	No
A	31	100	5	0	No	No
A	32	1	0	0	No	No
A	33			0	No	No
A	34	142	1	0	No	
A	35	78	2	0	Yes	No
A	36	600	200		No	
A	37	20	13	1	No	
A	38	45	15	0	No	
A	39	100	10	0	No	
A	40	80	6	2	Yes	No
A	41	250	25	0	Yes	No
A	42	4	0	0	No	
A	43	14	2	0	No	No
A	44	171	15	0	Yes	No
A	45	54	6	0	Yes	No
A	46	10	6	0	No	
A	47	0	0	0	No	No
A	48	24	0	0	No	
A	49	180	2	0	Yes	No
A	50					
A	51					

Hunting Guides

case		Number of clients Using			Recommend	
		Manual	Semiauto	AK47 et.al.	Any	AK47 et.al.
A	52	24	16	0	No	
A	53	600	100	12	No	
A	54	18	6	0	No	
A	55	0	0	0	No	
A	56	0	0	0	No	
A	57	40	4	0	No	
A	58					
A	59	40	10	0	No	No
A	60	60	2	0	No	No
A	61	63	4	0	Yes	No
A	62	40	4	0	No	
A	63	8	0	0	Yes	No
A	64	27	1	0	Yes	No
A	65	50	9	0	Yes	No
A	66	35	2	0	No	
A	67	6	0	0	Yes	No
A	68	6	3		No	
A	69	50	20	0	No	
A	70		0	0	Yes	No
A	71	27	1	0	Yes	
A	72	85	0	0	Yes	No
A	73	56	24	0	Yes	No
A	74	25	25	0	Yes	No
A	75	100	20	0	No	
A	76	50	15	3	No	
A	77	15	4	0	No	
A	78	12	0	0	Yes	No
A	79	75	0	0	No	
A	80					
A	81	0	0	0	No	
A	82	0	0	0	No	
A	83	12	4	0	No	No
A	84	40	0	0	Yes	No
A	85	24	0	0	No	
A	86	17	0	0	No	No
A	87	16	3	0	Yes	No
A	88	45	10	0	No	
A	89	11	7	7	Yes	Yes
A	90	35	1	0	Yes	No
A	91	25	2	0	Yes	No
A	92	0	0	0		
A	93	75	40	0	Yes	No
A	94	60	2	0	Yes	No
A	95	26	0	0	No	
A	96	20	0		No	No
A	97	65	11	0	Yes	No
A	98	40	5	0	Yes	No
A	99	26	5	0	No	
A	100	13	2	0	No	
A	101					
A	102	45	6	0	No	No

Hunting Guides

case		Number of clients Using			Recommend	
		Manual	Semiauto	AK47 et.al.	Any	AK47 et.al.
A	103	120	4	0	No	
A	104				Yes	
A	105	150	50	0	No	No
A	106	80	20	0	Yes	No
A	107	40	0	0	No	No
A	108	10	0	0	No	
A	109	160	40	0	Yes	No
A	110	10	10	0	No	No
A	111	6	0	0	No	
A	112					
A	113	150	150	100	Yes	Yes
A	114	50	25	0	No	No
A	115	19	0	0	Yes	No
A	116	80	3	0	No	
A	117	40	10	0	Yes	No
A	118					
A	119	50	0	0	Yes	No
A	120	0	0	0	No	
A	121	0	0	0		
A	122	120	15	0	Yes	No
A	123	10	0	0	Yes	No
A	124	22	0	0	Yes	No
A	125	40	40	20	No	
A	126	50	10	0	Yes	No
A	127	60	20	0	Yes	No
A	128	14	0	0	No	No
A	129	13	16	4	No	
A	130	80	4	0	Yes	No
A	131	12	2	0	Yes	No
A	132		4	0	Yes	No
A	133	50	26	7	No	No
A	134	12	0	0	No	
A	135	2	10	3	No	
A	136	2	1	1	Yes	No
A	137	28	0	0	Yes	No
A	138	45	10		No	
A	139	46	59	0	Yes	No
A	140			0	Yes	No
A	141	40	10	0	No	No
A	142	70	20	0	Yes	No
A	143	50	3	0	No	No
A	144	60	6	0	Yes	No
A	145	140	0	0	Yes	No
A	146	20	4	1	Yes	No
A	147	10	1	0	Yes	No
A	148	0	0	0	No	No
A	149	37	0	0	Yes	No
A	150			0	Yes	No
A	151	6	10	0	No	No
A	152	110	5	0	No	
A	153	15	17		Yes	No

case		Number of clients Using			Recommend	
		Manual	Semiauto	AK47 et.al.	Any	AK47 et.al.
A	154	18	4	0	No	
A	155	25	3	0	Yes	No
A	156	60	6	3	No	
A	157	20	0	0	No	
A	158	88	46	0	No	No
A	159	68	19	3	Yes	Yes
A	160	25	5	0	No	
A	161	15	0	0	No	
A	162	75	10	0	No	
B	1				No	
C	1	25	0	0	Yes	No
C	2	55	10	6	Yes	Yes
C	3	60	30	0	No	
C	4	80	20	0	No	
C	5	10	0	0	No	No
C	6	25	6	0	No	
C	7	66	10	1	No	
C	8	24	0	0	Yes	No
C	9	10	15	15	No	
C	10	35	15	9	Yes	Yes
C	11			0	No	
C	12					No
C	13	25	10	0	No	
C	14	60	20	0	Yes	No
C	15	20	0	0	Yes	No
C	16	14	0	0	No	
C	17		0	0	Yes	No
C	18	18	25	5	Yes	Yes
C	19	125	50	5	Yes	No
C	20	20	5	2	No	
C	21		0	0	Yes	No
C	22	30	0	0	No	No
C	23	150	20	0	Yes	No
C	24	60	0	0	No	
C	25	16	7	6	Yes	Yes
C	26	300	650	400	No	
C	27	20	15	8	Yes	Yes
C	28	3	5	2	No	
C	29	45	6	0	Yes	No
C	30				No	
C	31	30	0	0	Yes	No
C	32			0	Yes	No
C	33	35	4	0	Yes	No
C	34	25	5	0	Yes	No
C	35				Yes	No

Hunting Guides

case		Make	Other Make	Model	Caliber
A	1				
A	2				
A	3				
A	4	Browning		BAR	300
A	5				
A	6	Remington		742	30.06
A	7	Browning		BAR	30.06, .270, 7MM, 300 Mag
A	8				
A	9				
A	10				
A	11	Remington		740-7400	20, 30
A	12				
A	13	Remington		700	7 mm mag
A	14	Remington		7400	270
A	15				
A	16				
A	17				
A	18				
A	19	Browning			30.06
A	20	Remington		742	30.06
A	21				
A	22				
A	23	Browning		?	300 mag
A	24	Remington			30.06
A	25	Remington			30.06
A	26	Browning		BAR	30.06
A	27	Remington			30.06
A	28		?	?	06
A	29				
A	30				
A	31	Browning		automatics	
A	32				
A	33				
A	34	Remington			.3006
A	35	Browning			7 mm
A	36	Browning			30.06
A	37	Browning		BAR	30.06
A	38	Browning		br	7 mm, 300win, 30.06
A	39	Remington		7600	.270 win, .30-06, .280 rem
A	40	Browning		Bar mark II	300 win mag
A	41	Remington			
A	42				
A	43	Remington		7600	243 - 7 mm mag
A	44				30.06, 300 winmag, .338, 270
A	45	Browning		BAR Automatic	30.06

Q4. Three most commonly used rifles

			Q4. Three most commonly used rifles		
A	46	Browning		BAR	7 mm, 30.06
A	47				
A	48				
A	49				
A	50				
A	51				
A	52	Browning		BAR	7 mm mag/30.06
A	53	Browning		BAR	30.06, 300 wm
A	54	Browning		BAR	30.06
A	55				
A	56				
A	57	Browning		semi-auto	300 mag
A	58				
A	59				
A	60				
A	61	Browning			30.06
A	62	Browning			7 mm
A	63	Browning		BAR	.270 - 300 win mag
A	64	Browning		BAR	30.06
A	65	Browning		semi-auto	.308
A	66	Browning			
A	67				
A	68	Remington		7400	30.06
A	69	Browning			
A	70				
A	71	Browning		Not sure	
A	72				
A	73	Browning		BARR	30.06
A	74	Browning		BAR	300
A	75	Remington		7400 old 752	270 and 30.06
A	76	Browning		BAR	308, 30.06, 300win, 338 win
A	77	Remington			308
A	78	Browning			300, 270, 30.06
A	79				
A	80				
A	81				
A	82				
A	83				30 caliber or bigger for elk
A	84				
A	85				
A	86				
A	87	Browning			30.06 and 7 mm
A	88	Browning		BAR	7 mm, .300, .270
A	89	Other	Russian	SKS	7.62
A	90	Browning			1 or 2 in over 50 years
A	91	Browning			300 win mag

		Q4. Three most commonly used rifles			
A	92				
A	93				
A	94	Browning		BAR	
A	95				
A	96				
A	97	Browning		BAR	300-06-270
A	98	Browning			300, 30.06
A	99	Other	Savage		7 mm
A	100	Browning		?	7 mm mag
A	101				
A	102	Browning	Only 1 I recall	BAR	30.06
A	103				
A	104				
A	105				
A	106	Browning		BAR	300 win mag
A	107				
A	108				
A	109	Browning			30.06
A	110	Remington		700	30.06, 270, 7 mm
A	111				
A	112				
A	113	Other	Weatherby		300 mag
A	114	Browning			7 m mag
A	115				
A	116				
A	117	Browning			
A	118				
A	119				
A	120				
A	121				
A	122	Browning		U/K	.338 mag
A	123				
A	124				
A	125				
A	126	Remington		742	243, 30.06
A	127	Winchester		?	30.06
A	128	Winchester			270, 306
A	129	Browning		BAR	7 mm and 243
A	130	Browning			30.06
A	131	Browning		BAR	.7 mm mag
A	132	Remington			30.06
A	133			AK 47	223
A	134				
A	135	Remington			270
A	136	Browning		BAR	
A	137				

Q4. Three most commonly used rifles					
A	138	Winchester			30.06
A	139	Browning		BAR	270, 7 mm
A	140	Browning			7 mm
A	141				
A	142	Browning			7 mm mag
A	143				
A	144	Browning			30.06
A	145				
A	146	Browning		BDL	7mg
A	147	Browning		BAR	308
A	148				
A	149				
A	150	Remington			
A	151	Browning		BAR	308
A	152	Remington			various 270 - 338
A	153	Browning			30
A	154	Browning		BAR	7 mm mag
A	155				30.06
A	156	Other	BAR		
A	157				
A	158	Remington		280	280
A	159	Browning			7 mm mag
A	160	Remington		Semiauto	30.06
A	161				
A	162	Browning			30.06
B	1				.308, 30-06, .270
C	1				
C	2	Other	AK-47	Antelope Hunter	30
C	3	Browning		Auto	30.06
C	4	Browning		Bar	7mm
C	5				
C	6				
C	7	Browning			30.06
C	8				
C	9	Other	FN-FAL		308
C	10	Remington		742	30.06
C	11	Browning			306
C	12				
C	13	Remington			.06 - 7mm
C	14	Browning		BAR	7mm
C	15				
C	16				
C	17				
C	18	Ruger		Ranch Rifle	223
C	19	Other	AK47		
C	20	Browning		BAR	300 win mag

			Q4. Three most commonly used rifles		
C	21	Other	Bolt-action or pump		
C	22				
C	23	Browning			30.06
C	24				
C	25	Other	AK47		7.62-39
C	26	Other	HK	93	.308
C	27	Browning		BAR	7mm
C	28	Other	Norinco	SKS Type 56	7.62X39
C	29	Browning		BAR	30.06 -.300
C	30				
C	31				
C	32	Browning			3.06 - 7mm
C	33	Remington			30.06
C	34	Remington		741	.270 - 30.06
C	35	Remington			.270
A	1				
A	2				
A	3				
A	4	Remington		7400	30.06
A	5				
A	6	Browning			30.06
A	7	Remington		700	30.03, 270, 7 mm
A	8				
A	9				
A	10				
A	11	Winchester		100	30
A	12				
A	13	Winchester		70	300 mag
A	14	Remington		7400	30.06
A	15				
A	16				
A	17				
A	18				
A	19	Remington		7400	30.06
A	20	Browning			7 mm mag
A	21				
A	22				
A	23				
A	24	Browning			30.06
A	25	Browning			30.03 to 300 mag
A	26	Remington		Fieldmaster	30.06
A	27				
A	28				
A	29				
A	30				
A	31	Remington		automatics	

		Q4. Three most commonly used rifles		
A	32				
A	33				
A	34				
A	35				
A	36	Remington			270 - 30.06
A	37	Remington		7400	30.06
A	38				
A	39	Browning		BAR	.270 win, 7 mm mag
A	40	Remington		7400	30.06
A	41	Browning			
A	42				
A	43	Browning		BAR	243 - 7 mm mag
A	44				
A	45				
A	46	Remington		1100	12 gauge
A	47				
A	48				
A	49				
A	50				
A	51				
A	52	Remington		7400	30.06
A	53	Remington		7400/742	30.06
A	54				
A	55				
A	56				
A	57	Remington		semi-auto	30.06
A	58				
A	59				
A	60				
A	61	Other	Savage		7 mm mag
A	62	Remington			30.06
A	63	Remington		742	.270 - 30.06
A	64				
A	65	Winchester		semi-auto	.308
A	66	Remington			
A	67				
A	68	Remington		7400	.308
A	69	Remington			
A	70				
A	71	Remington		742	30.06
A	72				
A	73	Remington			30.06
A	74	Remington		?600	30.06
A	75	Browning		BAR	270/338 and 30.06
A	76	Other	AK-47		30
A	77	Remington			30.06

		Q4. Three most commonly used rifles			
A	78	Remington			?	300, 270, 30.06
A	79				
A	80				
A	81				
A	82				
A	83				
A	84				
A	85				
A	86				
A	87	Remington			30.06
A	88	Remington		742, 7400	30.06. .270
A	89	Other	Heckler-Koch	HK91	308
A	90	Remington			
A	91	Remington			30.06
A	92				
A	93				
A	94				
A	95				
A	96				
A	97				
A	98	Remington		760	.300, 30.06, 270
A	99	Browning			7 mm
A	100	Remington		742	30.06
A	101				
A	102				
A	103				
A	104				
A	105				
A	106				
A	107				
A	108				
A	109	Winchester			308
A	110				
A	111				
A	112				
A	113	Remington		700	7 mm mag
A	114	Remington		742 Wingmaster	30.06
A	115				
A	116				
A	117	Remington			
A	118				
A	119				
A	120				
A	121				
A	122				
A	123				

		Q4. Three most commonly used rifles		
A	124			
A	125			
A	126	Ruger		22
A	127	Marlin		?	.308
A	128	Remington			7 m
A	129			
A	130			
A	131	Browning		BAR	30.06
A	132			
A	133	Ruger		Mini 14	223
A	134			
A	135	Remington			243
A	136	Other	HK 91	
A	137			
A	138	Browning			308
A	139	Remington		742	30.06 - 6 mm
A	140	Remington			30.06
A	141			
A	142	Browning			300 win mag
A	143			
A	144	Browning			7 mm mag
A	145			
A	146	Browning		BDL	300
A	147			
A	148			
A	149			
A	150	Winchester		
A	151	Remington		742	30.06
A	152	Ruger			various 270 - 338
A	153	Winchester			30
A	154	Browning		BAR	30.06
A	155			
A	156	Other	AK-47	
A	157			
A	158	Winchester			338
A	159	Remington			30.06
A	160			
A	161			
A	162	Remington		742	30.06, 270
B	1			
C	1			
C	2			
C	3	Winchester		Auto	30.06
C	4	Browning		Bar	338
C	5			
C	6			

		Q4. Three most commonly used rifles			
C	7	Remington			30.06
C	8				
C	9	Other	Uzi		9mm
C	10	Other	AK-47	Hunter	7.62x39
C	11	Other	Weatherby		300
C	12				
C	13	Winchester			.06 - 7mm
C	14	Browning			300
C	15				
C	16				
C	17				
C	18	Other	AK-47		
C	19	SigArms		550-1	
C	20	Ruger		Mini 14	.223
C	21				
C	22				
C	23	Remington		742	30.06
C	24				
C	25	Other	MAK-90		7.62-39
C	26	Other	HK	91	0.223
C	27	Remington		7400 Series	30.06
C	28	Remington		7600	30.06
C	29	Remington		742	.308 - 3.06
C	30				
C	31				
C	32	Remington			30.06 - 7mm
C	33	Browning			300 win
C	34	Browning			.270 - 30.06
C	35	Browning			300
A	1				
A	2				
A	3				
A	4	Ruger		Mini 14	223
A	5				
A	6	Other	Savage		270
A	7				
A	8				
A	9				
A	10				
A	11				
A	12				
A	13	Browning		A-bolt	270
A	14				
A	15				
A	16				
A	17				

Q4. Three most commonly used rifles				
A	18				
A	19				
A	20				
A	21				
A	22				
A	23				
A	24				
A	25				
A	26	Other	China	SKS	7.62x37
A	27				
A	28				
A	29				
A	30				
A	31				
A	32				
A	33				
A	34				
A	35				
A	36	Winchester			270 - 30.06
A	37				
A	38				
A	39				
A	40	Ruger			44 mag
A	41				
A	42				
A	43	Ruger			223 - 30.06
A	44				
A	45				
A	46				
A	47				
A	48				
A	49				
A	50				
A	51				
A	52				
A	53	Ruger		Mini-14	.223
A	54				
A	55				
A	56				
A	57	Ruger		semi-auto	35 cal
A	58				
A	59				
A	60				
A	61				
A	62	Ruger		Mini 14	223
A	63				

Q4. Three most commonly used rifles				
A	64				
A	65				
A	66				
A	67				
A	68				
A	69				
A	70				
A	71				
A	72				
A	73				
A	74	Browning		BAR	30.06
A	75				
A	76	Remington			30.06, 270
A	77	Browning			300
A	78				
A	79				
A	80				
A	81				
A	82				
A	83				
A	84				
A	85				
A	86				
A	87				
A	88				
A	89	Other	Springfield Armory	FNG	308
A	90				
A	91				
A	92				
A	93				
A	94				
A	95				
A	96				
A	97				
A	98				
A	99				
A	100				
A	101				
A	102				
A	103				
A	104				
A	105				
A	106				
A	107				
A	108				
A	109				

		Q4. Three most commonly used rifles			
A	110				
A	111				
A	112				
A	113	Other	All		30.06
A	114	Remington		721	270
A	115				
A	116				
A	117				
A	118				
A	119				
A	120				
A	121				
A	122				
A	123				
A	124				
A	125				
A	126	Browning	Remington	Shotguns	12 gauge
A	127	Remington			.308 or 30.06
A	128	Other	Savage		308
A	129				
A	130				
A	131				
A	132				
A	133	Browning		BAR	7 mm
A	134				
A	135	Browning		742	30.06
A	136	Other	AK 47		
A	137				
A	138				
A	139	Other	Weatherby		300 m
A	140				
A	141				
A	142				
A	143				
A	144				
A	145				
A	146	Ruger		#1	7 mag
A	147				
A	148				
A	149				
A	150	Browning			
A	151				
A	152	Browning			various 270 - 338
A	153				
A	154	Browning		BAR	8 mm mag
A	155				

		Q4. Three most commonly used rifles			
A	156	Other	Uzi		
A	157				
A	158	Browning			300
A	159				
A	160				
A	161				
A	162				
B	1				
C	1				
C	2				
C	3	Browning		Auto	270
C	4	Browning		Bar	300
C	5				
C	6				
C	7				
C	8				
C	9	Other	HK91		
C	10	Browning		BAR	30.06
C	11				
C	12				
C	13	Browning			300
C	14				
C	15				
C	16				
C	17				
C	18				
C	19				
C	20	Other	AK47		7.62 x 39
C	21				
C	22				
C	23	Remington		742	308, 270
C	24				
C	25		M1-A1		.223
C	26				
C	27	Winchester	Various	M1 Garand	30.06
C	28				
C	29			M1A1	30.06
C	30				
C	31				
C	32				
C	33				
C	34				
C	35				

	Q 6. Rifles recommended for clients			
case	Make	Other Make	Model	Caliber
A 1				
A 2	Ruger			30.06
A 3 '				
A 4	Other	Weatherby	Mark V	300
A 5				30.06
A 6				
A 7				
A 8				
A 9				
A 10				
A 11				
A 12				
A 13				
A 14				
A 15				
A 16				
A 17				
A 18				
A 19				
A 20				
A 21	Winchester			30.06, .270
A 22	Remington		700	7 mm or larger
A 23	Winchester		70	25 to 30
A 24	Remington		710	30.06
A 25		Any make	Bolt action	Does not recommend
A 26	Winchester		70	30.06 or larger
A 27	Other	Weatherby		300
A 28	Other	bolt action		270 and up
A 29				
A 30		hunter's choice		.270
A 31				
A 32				
A 33				
A 34				
A 35	Winchester		70	300 win mag
A 36				
A 37				
A 38				
A 39				
A 40	Remington			30.06 - 300 win mag
A 41				
A 42				
A 43				
A 44				30.06, 300winmag, 338, 270
A 45	Browning		Bolt Action	25.06 - 328

	Q 6. Rifles recommended for clients			
case	Make	Other Make	Model	Caliber
A 46				
A 47				
A 48				
A 49	Other	Weatherby		300 mag
A 50				
A 51				
A 52				
A 53				
A 54				
A 55				
A 56				
A 57				
A 58				
A 59				
A 60				
A 61	Remington		Bolt Action	300 mag
A 62				
A 63	Other	bolt action repeating rifles		30.06 to .338 winmag
A 64	Winchester		70	338
A 65	Remington		bolt action	308,25-06,243,7 mm mag,30.06,22-250,300 mag all
A 66				
A 67	Ruger		#1	7 mm, 30.06, 7 mm mag
A 68				
A 69				
A 70	Other		Bolt Action	30.06
A 71				300 mag
A 72	Other	Any make	Any model	7 mm, 270, 30.06, 25.06
A 73				
A 74	Browning		BAR	300 win mag
A 75				
A 76				
A 77				
A 78	Browning		Bolt action	
A 79				
A 80				
A 81				
A 82				
A 83				
A 84				
A 85				
A 86				
A 87	Remington		700	30.06, 7 mm, 270
A 88				
A 89	Other	Russian	SKS	7.62
A 90	Other	Weatherby		7 mm mag

	case	Make	Other Make	Model	Caliber
	Q 6. Rifles recommended for clients				
A	91	Remington		700	7 mag
A	92				
A	93	Winchester		70	300 mag
A	94	Other	Any bolt action		270 or larger
A	95				
A	96				
A	97	Other	Any bolt action		30 or larger, on semiauto same
A	98				
A	99				
A	100				
A	101				
A	102				
A	103				
A	104				
A	105				
A	106	Other	Weatherby		300 magnum
A	107				
A	108				
A	109	Remington		70	7 mm
A	110				
A	111				
A	112				
A	113				
A	114				
A	115				
A	116				
A	117				magnum
A	118				
A	119	Remington		700	7 mm
A	120				
A	121				
A	122				
A	123				
A	124				
A	125				
A	126				300 mag, 338 mag, 30.06
A	127				
A	128				
A	129				
A	130	Remington		700	7 mm magnum
A	131				
A	132	Other	Weatherby		300 mag
A	133				
A	134				
A	135				

		Q 6. Rifles recommended for clients		
case	Make	Other Make	Model	Caliber
A	136			
A	137	Remington		700	7 mm
A	138			
A	139	Browning		BAR	7 m or 270
A	140			
A	141			
A	142				30.06
A	143			
A	144	Browning			from 7 mm mag to 338 mag for deer and elk
A	145	Winchester			30.06
A	146	Browning		BDL	7 mag
A	147	Remington		700 BDL	7 mm
A	148			
A	149			
A	150	Browning		Bolt action	
A	151			
A	152			
A	153	Remington		700	30
A	154			
A	155	Other	Weatherby		300
A	156			
A	157			
A	158			
A	159	Browning	Ruger		243, 30.06, 7 mm mag, 340 weather, .338
A	160			
A	161			
A	162			
B	1				7.62 x 39
C	1	Other	Manually operated		
C	2	Ruger		77	300
C	3			
C	4			
C	5			
C	6			
C	7			
C	8	Remington		700	270
C	9			
C	10	Other	HK	91	.308
C	11			
C	12			
C	13			
C	14	Other	Bolt-action w/ belted mag		Calibers, make and model mean nothing
C	15	Other	Bolt-action		30.06-7mm
C	16			
C	17	Other	Bolt-action		

Q 6. Rifles recommended for clients				
case	Make	Other Make	Model	Caliber
C 18	Ruger		Ranch Rifle	223
C 19				.243 and larger
C 20				
C 21				
C 22				
C 23	Other	Bolt-action		7mm mag
C 24				
C 25	Other	Savage		7mm mag
C 26				
C 27	Winchester		70	30.06
C 28				
C 29	Winchester		70	30.06 - .338
C 30				
C 31	Winchester		Manual, bolt	300
C 32	Remington		All	270 - 7mm
C 33	Winchester		70	30.06 - .300 win
C 34	Other	Bolt-action		270 or larger for elk and deer
C 35	Other	Bolt-action or semiautos		.270 or larger
A 1				
A 2	Remington			7 mm
A 3				
A 4	Winchester		70	300
A 5				
A 6				
A 7				
A 8				
A 9				
A 10				
A 11				
A 12				
A 13				
A 14				
A 15				
A 16				
A 17				
A 18				
A 19				
A 20				
A 21	Remington		70	30.06
A 22	Winchester		70	7 mm or larger
A 23	Remington		700	25 to 30
A 24	Remington			300 Mag
A 25				
A 26	Browning		A bolt	30.06 or larger
A 27				300 win mag, 30.06 or 270

	case	Make	Other Make	Model	Caliber
A	28				
A	29				
A	30		hunter's choice		.308
A	31				
A	32				
A	33				
A	34				
A	35	Remington		700 BDL	7 mm
A	36				
A	37				
A	38				
A	39				
A	40	Winchester			30.06 - 300 win mag
A	41				
A	42				
A	43				
A	44				
A	45	Remington		Bolt Action	25.06 - 328
A	46				
A	47				
A	48				
A	49				
A	50				
A	51				
A	52				
A	53				
A	54				
A	55				
A	56				
A	57				
A	58				
A	59				
A	60				
A	61	Other	Savage	Bolt Action	7 mm mag
A	62				
A	63				
A	64	Remington		700	300 win mag
A	65	Other	Weatherby		
A	66				
A	67	Remington		Bolt Action	7 mm, 30.06, 7 mm mag
A	68				
A	69				
A	70			Pump	30.06
A	71				7 mm mag
A	72				

Q 6. Rifles recommended for clients

case		Make	Other Make	Model	Caliber
A	73				
A	74	Winchester		7C	300 win mag
A	75				
A	76				
A	77				
A	78	Remington		Bolt Action	
A	79				
A	80				
A	81				
A	82				
A	83				
A	84				
A	85				
A	86				
A	87	Browning			308, 7 mm, 30.06
A	88				
A	89	Other	Heckler-Koch	HK-91	308
A	90				
A	91	Winchester		70	300 mag
A	92				
A	93	Browning		Mark II	300 mag, 280-270-25.06
A	94				
A	95				
A	96				
A	97	Other	Semi-auto		30 cal or larger
A	98				
A	99				
A	100				
A	101				
A	102				
A	103				
A	104				
A	105				
A	106	Remington		700	300 win mag
A	107				
A	108				
A	109	Winchester			300 mag, 30.06
A	110				
A	111				
A	112				
A	113				
A	114				
A	115				
A	116				
A	117				

	case	Make	Other Make	Model	Caliber
			Q 6. Rifles recommended for clients		
A	118				
A	119	Other	Weatherby		300
A	120				
A	121				
A	122				
A	123				
A	124				
A	125				
A	126				
A	127				
A	128				
A	129				
A	130				
A	131				
A	132	Other	Weatherby		700 mag
A	133				
A	134				
A	135				
A	136				
A	137	Other	Weatherby		300
A	138				
A	139	Remington		742	30.06 or 6 mm
A	140				
A	141				
A	142				7 mm recommended for deer and elk
A	143				
A	144	Other	Weatherby		from 7 mm mag to 338 for deer
A	145	Other	Weatherby		300
A	146	Browning		BDC	300
A	147				
A	148				
A	149				
A	150	Winchester		Bolt Action	
A	151				
A	152				
A	153	Remington		700	7 mm
A	154				
A	155	Other	Weatherby		7 mm
A	156				
A	157				
A	158				
A	159	Winchester	Remington		340 Weather - .338 mag
A	160				
A	161				
A	162				

		Q 6. Rifles recommended for clients			
case		Make	Other Make	Model	Caliber
B	1				
C	1				
C	2	Browning			300
C	3				
C	4				
C	5				
C	6				
C	7				
C	8	Remington		700	280
C	9				
C	10	Winchester		70	.270
C	11				
C	12				
C	13				
C	14				
C	15				
C	16				
C	17	Other	Pump		
C	18	Other	AK-47		
C	19				6mm
C	20				
C	21				
C	22				
C	23	Other	Bolt-action		.30
C	24				
C	25	Other	Bolt-action		30.06
C	26				
C	27	Ruger		77	.300 win mag
C	28				
C	29	Remington		700	30.06-.338
C	30				
C	31	Remington		Manual bolt	300
C	32	Browning		All	.270 - 7mm
C	33	Ruger		77	30.06 - .300 win
C	34				
C	35				
A	1				
A	2	Winchester			376
A	3				
A	4	Winchester		70	270
A	5				
A	6				
A	7				
A	8				
A	9				

		Q 6. Rifles recommended for clients			
case		Make	Other Make	Model	Caliber
A	10				
A	11				
A	12				
A	13				
A	14				
A	15				
A	16				
A	17				
A	18				
A	19				
A	20				
A	21	Remington		70	.270
A	22				
A	23	Other	Any bolt action	1-5 shotmag	25 to 30
A	24	Other	Weatherby		300 mag
A	25				
A	26				
A	27				
A	28				
A	29				
A	30				
A	31				
A	32				
A	33				
A	34				
A	35				
A	36				
A	37				
A	38				
A	39				
A	40	Ruger			30.06 - 300 win mag
A	41				
A	42				
A	43				
A	44				
A	45	Winchester		Bolt Action	25.06 - 328
A	46				
A	47				
A	48				
A	49				
A	50				
A	51				
A	52				
A	53				
A	54				

Q 6. Rifles recommended for clients				
case	Make	Other Make	Model	Caliber
A 55				
A 56				
A 57				
A 58				
A 59				
A 60				
A 61	Other	Weatherby	Bolt Action	338 mag
A 62				
A 63				
A 64	Other	Weatherby Mark V		300 Wea Mag
A 65	Winchester	Browning		
A 66				
A 67	Winchester	Bolt Action		
A 68				
A 69				
A 70			Bolt Action	7 mm
A 71				
A 72				
A 73				
A 74	Browning		A Bolt	300 win mag
A 75				
A 76				
A 77				
A 78				
A 79				
A 80				
A 81				
A 82				
A 83				
A 84				
A 85				
A 86				
A 87	Other	Weatherby		300, 7 mm, 338
A 88				
A 89	Other	Springfield Armory	FNG	308
A 90				
A 91	Ruger		77	300 mag
A 92				
A 93	Ruger		M77	270, 26-06, 300 mag
A 94				
A 95				
A 96				
A 97				
A 98				
A 99				

Q 6. Rifles recommended for clients				
case	Make	Other Make	Model	Caliber
A	100				
A	101				
A	102				
A	103				
A	104				
A	105				
A	106	Browning		1895	45-70 govt
A	107				
A	108				
A	109				
A	110				
A	111				
A	112				
A	113				
A	114				
A	115				
A	116				
A	117				
A	118				
A	119	Other	Savage		270 or 30.06
A	120				
A	121				
A	122				
A	123				
A	124				
A	125				
A	126				
A	127				
A	128				
A	129				
A	130				
A	131				
A	132				
A	133				
A	134				
A	135				
A	136				
A	137				
A	138				
A	139				
A	140				
A	141				
A	142				300 winmag recommended
A	143				
A	144	Remington	Weatherby		from 270 to 338 for deer and elk

Q 6. Rifles recommended for clients				
case	Make	Other Make	Model	Caliber
A 145	Remington			270
A 146	Ruger		#1	7 mag
A 147				
A 148				
A 149				
A 150				All bolt action with a round nose point
A 151				
A 152				
A 153				
A 154				
A 155				
A 156				
A 157				
A 158				
A 159				300mag,416Rigby,375mag,270 mag,500 nitroxpress
A 160				
A 161				
A 162				
B 1				
C 1				
C 2	Other	Sako		300
C 3				
C 4				
C 5				
C 6				
C 7				
C 8				
C 9				
C 10	Winchester		100	.308
C 11				
C 12				
C 13				
C 14				
C 15				
C 16				
C 17	Other	Weatherby		243 to 300
C 18				
C 10				
C 20				
C 21				
C 22				
C 23				
C 24				
C 25				
C 26				

Q 6. Rifles recommended for clients				
case	Make	Other Make	Model	Caliber
C 27	Springfield		M Garard	30.06 - 308
C 28				
C 29	Browning		A bolt	30.06 - .338
C 30				
C 31				
C 32	Ruger		All	.270 - 7 mm
C 33	Browning		A bolt	30.06 - 300 win
C 34				
C 35				

case		Make	Other Make	Model	Caliber
A	26	AK47			7.62x37
A	89	Other	Russian	SKS	7.62
A	113	FN-FAL			
A	159	AK47			
C	2	AK47		Antelope and Varmints and Target Shooters	30
C	10	AK47			7.62x39
C	18	AK47			
C	25	AK47			7.62
C	27	FN-FAL			308
A	26		SKS		7.62x37
A	89	HK91			308
A	113		HK 99		
C	2	AK47		Antelope and Varmints and Target Shooters	243
C	10	HK91			308
C	25		MAK 90		7.62
C	27		Century	L1A1	308
A	89	Other	Springfield Armory	FNG	308
A	113	HK93			
C	10	HK93			223
C	25		M-15		223
C	27	HK91	And clones		308

Q 8. Recommended rifles based on AK47 et.al.

Additional Comments by Hunting Guides

Additional comments:

(8) The respondent answered questions 1, 2, 3, and 5 with "None of your business." He then stated in question 4: "It's none of your business what kind, make, model or how many guns law abiding citizens of the U.S. own, prefer to shoot."

(9) The respondent wrote that he was no longer in business but that he had owned a waterfowl operation and upland bird operation (shotguns only). He added that assault rifles were not true sporting rifles and that they should be limited to use by the military and law enforcement agencies. However, he felt that true sporting weapons that can be modified into some "quasi-assault weapons" should not be restricted. He stated that he supported the effort to get military weapons off the streets but did not want the rights of true sportsmen to be affected.

(10) Although licensed, the respondent did not guide anyone during the past year.

(11) The respondent stated in question 6 that he recommends any legal caliber rifle that client is comfortable with and that is capable of killing the desired game.

(12) For question 6, the respondent replied that he didn't recommend any specific make or model, other than whatever his clients are most comfortable using so long as the weapons are legal for the particular game.

(15) The respondent stated that his organization was solely recreational wildlife watching and photography.

(17) The respondent did not answer the questions but informed us that it is illegal in Hawaii to hunt turkey with a rifle.

(23) The respondent stated that the study rifles were more suitable for militants than sportsmen. He added, "If they want to use these weapons let them go back to the service and use them to defend our country, not against it."

(25) The respondent stated that, in his 35 years of conducting big game hunts, he had never seen any of the study rifles used for hunting. He suggested that the rifles are made to kill people, not big game.

(26) The respondent recommended bolt-action rifles for his clients but stated that he doesn't demand that they use such rifles. The respondent recommended the study rifles in close-range situations in which there are multiple targets that may pose a danger to the hunter (e.g., coyotes, foxes, mountain lions, and bears).

(27) The respondent stated that he recommended the study rifles for hunting but not any specific make.

(32)   The respondent said that most of his clients are bow or pistol hunters. He said that there is little if any use for the study rifles in his outfitting service because it focuses on hunts of mountain lions and bighorn sheep. However, he did recommend the study rifles on target ranges and in competitive shooting situations and cited his right to bear arms.

(35)   The respondent recommended bolt-action rifles for his clients.

(40)   The respondent stated that semiautomatic rifles (such as the AK47) and others are useful for predator hunting.

(41)   The respondent said that he recommended only ranges of calibers deemed suitable but not makes and models of specific rifles.

(44)   The respondent recommended the following calibers for hunting without any specific makes or models: 30.06, 300 Win mag, 338, and 270.

(47)   The respondent stated: "You are asking questions about certain makes of assault rifles, but you are going to end up going after ALL semiautomatic guns. I've spent about 21 years HUNTING with shotguns and I've used semiautomatic models. If you go down the list of times that one new law didn't end up being a whole sloo [sic] of other laws I would be surprised. Maybe some face-to-face with these weapons would be a good thing for politicians. If they see how they are used in 'the Real World' then they may make better amendments."

(49)   The respondent specifically recommended the study rifles only for grizzly bears or moose.

(50)   The respondent stated that his business involved waterfowl hunting, which uses only shotguns.

(51)   The respondent replied: "It is my opinion this is a one sided survey, and does not tell the real meaning and purpose of the survey. And that is to ban all sporting arms in the future. The way this survey is presented is out of line."

(53)   The respondent stated: "I recommend to all my hunters that they join the NRA, vote Republican, and buy a good semi-auto for personal defense."

(57)   The respondent stated that most of his clients use bolt-action rifles. He suggested that semiautomatics are not as accurate as bolt-action rifles.

(58)   The respondent stated that the survey did not pertain to his waterfowl hunting business since only shotguns are used. He added that he did not believe semiautomatics in general present any more threat to the public than other weapons or firearms. However, he suggested that cheaply made assault-type rifles imported from China and other countries are inaccurate and not suitable for hunting.

(59)   The respondent stated that he had no knowledge of the semiautomatic rifles beyond 30.06 or similar calibers for hunting. He added that he did not have a use for "automatic" weapons.

(64) The respondent stated: "We need to look at weapons and determine what the designer's _intent_ was for the weapon. We really _don't_ need combat weapons in the hunting environment. I personally would refuse to guide for anyone carrying such a weapon."

(65) The respondent recommended the following calibers for hunting: 7mm, 30.06, .308, .708, 25.06, .243, 22.250, and 300 mag. However, he stated that the study rifles are of no use to the sporting or hunting community whatsoever.

(71) The respondent stated that he mainly hunts elk but did not recommend any additional information about specific firearms except for using 300 mag and 7 mm mag calibers.

(73) The respondent recommended any bolt-action or semiautomatic in the 30 or 7mm calibers. However, he stated that he doesn't allow his clients to use any models based on assault rifles: "They are not needed for hunting. A good hunter does not have these."

(78) The respondent recommended bolt-action rifles for hunting, particularly Browning and Remington.

(80) Although the respondent stated that he does not conduct guides, he did not see a reason to allow any rifles other those manufactured specifically for hunting and sport shooting: "All assault rifles are for fighting war and killing humans."

(82) The respondent stated that he used shotguns only.

(84) The respondent said that he did not allow semiautomatic or automatic rifles in his business. He specifically recommended manually operated rifles.

(90) The respondent stated that all the semiautomatics like AK47s are absolutely worthless and that he found no redeeming hunting value in any AK47 type of rifle. He further explained that the purpose of hunting is to use the minimum number of shells, not the maximum: "I have only known 1 [person] in 50 years to use an AK47. He shot the deer about 30 times. That wasn't hunting, it was murder." He suggested that he would be willing to testify in Congress against such weapons.

(92) The respondent stated that he had been contacted in error, as he was not in the hunting guide business.

(98) The respondent recommended any rifle that a client can shoot the best.

(101) The respondent wrote a letter saying that his business was too new to provide us with useful information about client use; however, he stated that the Chinese AK47 does a proficient job on deer and similar sizes of game and may be the only rifle that some poor people could afford. He said that he is willing to testify to Congress about the outrageous price of certain weapons.

(102) The respondent did not recommend rifles but recommended calibers .270, 30.06, .300, and 7mm.

3

(103) The respondent stated that he had clients who used semiautomatic rifles, but he didn't know which makes or models.

(104) The respondent recommended any legal weapons capable of killing game, "including the types mentioned under the 2nd amendment."

(105) The respondent stated that the semiautomatic rifles used by his clients were Remingtons.

(112) The respondent stated that he could not provide any useful information because his business was too new.

(113) The respondent recommended whatever is available to knock down an elk. He recommended specific calibers: 30.06, 300, or 338.

(115) The respondent questioned why anyone would use a semiautomatic firearm to hunt game: "Anyone using such horrible arms should be shot with one themselves. Any big game animal does not have a chance with a rifle and now you say people can use semiautomatic rifles."

(116) The respondent had had three clients who used semiautomatics with 30.06 and 270-caliber ammunition; however, he didn't know the makes or models.

(118) The survey questions were not answered, but the respondent wrote: "This is a stupid survey. No one contends they hunt much for big game with an AK47. The debate is over the right to own one, which the 2nd amendment says we can."

(119) The respondent recommended bolt-action rifles for hunting.

(121) The respondent stated that he uses only shotguns in his operation.

(122) The respondent recommended rifles with the calibers of .270 - 30.06 or larger to the .300 mag or .338 mag. However, he said that anything other than a standard semiautomatic sporting rifle is illegal in Colorado, where his business is conducted.

(123) The respondent, who is a bighorn sheep outfitter, stated that the semiautomatic rifles have no place in big game hunting. He recommended basic hunting rifles with calibers of 270 or 30.06.

(124) The respondent, who hunts mainly deer and elk, recommended calibers 270, 30.06, 300 mag, 7mm, 8mm, or 338.

(125) The respondent said that his clients did use semiautomatics, but he didn't have any specific information about which ones.

(126) The respondent stated that the study rifles should remain in one's home or on private property. He would like to have some for personal use but would not recommend them for hunting. He further expressed his displeasure with the Brady bill and stated that criminals need to be held accountable for their actions.

(127) The respondent, who hunts mostly elk and deer, said that the AK47 is not powerful enough to hunt elk; however, it may be ideal for smaller game, like deer or antelope. He recommended any rifles of 30.06 caliber or larger for hunting.

(131) The respondent recommended bolt-action rifles for his clients with calibers .24, .25, 7 mm, or .30. He cited his preference because of fewer moving parts, their ease to fix, and their lack of sensitivity to weather conditions in the field. He added, however, that he had seen the study rifles used with good success.

(132) The respondent stated that the study rifles are not worth anything in cold weather.

(133) The respondent recommended handguns for hunting in calibers 41 or 44 mag.

(136) The respondent did not recommend any rifles by make, but he did recommend a caliber of .308 or larger for elk.

(140) The respondent recommended any good bolt or semiautomatic in 270 caliber and up. He added: "I feel the government is too involved in our lives and seek too much control over the people of our country. I am 65 yrs old and see more of our freedom lost every day. I believe in our country but I have little faith in [organizations] like the A.T.F."

(145) The responded stated: "Don't send these guns out west. Thanks!"

(148) The respondent did not hunt turkey or deer and had no additional information to provide.

(149) The respondent said that he recommends specific rifles to his clients if they ask, usually 270 to 7mm caliber big game rifles.

(150) The respondent recommended Winchester, Remington, or any other autoloading hunting rifle.

(152) The respondent said that he recommended caliber sizes but not specific rifles.

(159) The respondent recommended any gun with which a client can hit a target. He stated that the AK47 could be used for hunting and target shooting.

(174) The respondent recommended bolt-action rifles to his clients.

(175) The respondent said that most of his deer-hunting clients use bolt-action rifles, such as Rugers and Remingtons, in calibers of 30.06, 270, or 243. In his duck guide service, only shotguns are used.

(180) The respondent wrote: "We agree people should not be allowed to have semiautomatics and automatics. This does not mean that you silly bastards in Washington need to push complete or all gun control."

(182) The respondent felt that the survey is biased because it didn't ask about hunting varmints. He stated that many of the study rifles are suitable for such activity.

(184) The respondent did not recommend single shots or automatics and only allows bolt action or pumps for use by his clients.

(188) The respondent wrote that the study guns are good for small game hunting: "I have very good luck with them as they are small, easy to handle, fast-shooting and flat firing guns."

(192) The respondent submitted a letter with the survey: "I do not recommend the use of semiautomatic weapons for hunting in my area. Most of these weapons are prone to be unreliable because the owner does not know how to properly care for them in adverse weather. The FN-FAL, HK91, HK93, and SIG SG550-1 are excellent and expensive weapons very much suited to competition shooting.

"Have you surveyed the criminal element on their choice of weapons? I suspect the criminal use of the six weapons you mentioned do law-abiding citizens compare a very small percentage to the same weapon used. I realize that even one wrongful death is too many but now can you justify the over 300,000 deaths per year from government supported tobacco?

"Gun control does not work - it never has and it never will. What we need are police that capture criminals and a court system with the fortitude to punish them for their crimes."

(198) The respondent stated that this was his first year in and that it was mainly a bow-hunting business.

DEC 1 0 1997

O:F:S:DMS
3310

Dear Sir or Madam:

On November 14, 1997, the President and the Secretary
of the Treasury decided to conduct a review to
determine whether modified semiautomatic assault rifles
are properly importable under Federal law. Under
18 U.S.C. section 925(d)(3), firearms may be imported
into the United States only if they are determined to
be of a type generally recognized as particularly
suitable for or readily adaptable to sporting purposes.
The firearms in question are semiautomatic rifles based
on the AK47, FN-FAL, HK91, HK93, SIG SG550-1, and Uzi
designs.

As part of the review, the Bureau of Alcohol, Tobacco
and Firearms (ATF) is interested in receiving
information that shows whether any or all of the above
types of semiautomatic rifles are particularly suitable
for or readily adaptable to hunting or organized
competitive target shooting. We are asking that your
organization voluntarily complete the enclosed survey
to assist us in gathering this information. We
anticipate that the survey will take approximately
15 minutes to complete.

Responses must be received no later than 30 days
following the date of this letter; those received after
that date cannot be included in the review. Responses
should be forwarded to the Bureau of Alcohol, Tobacco
and Firearms, Department HSE, P.O. Box 50860,
Washington, DC 20091. We appreciate any information
you care to provide.

Sincerely yours,

John W. Magaw
Director

Enclosure

ATF SURVEY OF HUNTING/SHOOTING EDITORS
FOR RIFLE USAGE

1. Does your publication recommend specific types of centerfire semiautomatic rifles for use in **hunting medium game (for example, turkey) or larger game (for example, deer)?**

_____Yes *(Continue)* _____No *(Skip to #3)*

2. If your answer to item 1 is "Yes", please identify the specific centerfire semiautomatic rifles you recommend.

Make   Model   Caliber

_____

_____

_____

3. Does your publication recommend **against** the use of any semiautomatic rifles whose design is based on the **AK 47, FN-FAL, HK91, HK93, SIG 550-1, or Uzi** for use in **hunting medium game (for example, turkey) or larger game (for example, deer)?**

_____Yes *(Continue)* _____No *(Skip to #5)*

_____Yes, in certain circumstances. Please explain _____

_____*(Continue)*

4. If your answer to item 3 is "Yes" or "Yes, in certain circumstances", please identify the specific rifles that you recommend **against** using for **hunting medium game (for example, turkey) or larger game (for example, deer)?**

Make   Model   Caliber

_____

_____

_____

5. Does your publication recommend specific types of centerfire semiautomatic rifles for use in **high-power rifle competition**?

_____Yes *(Continue)* _____No *(Skip to #7)*

An agency may not conduct or sponsor, and a person is not required to respond to, the collection of information unless it displays a currently valid OMB control number.

OMB No. 1512-0542

6. If your answer to item 5 is "Yes", please identify the specific centerfire semiautomatic rifles you recommend.

<u>Make</u>   <u>Model</u>   <u>Caliber</u>

_____

_____

_____

7. Does your publication recommend **against** the use of any semiautomatic rifles whose design is based on the **AK 47, FN-FAL, HK91, HK93, SIG 550-1, or Uzi** for use in **high-power rifle competition**?

_____Yes *(Continue)* _____No *(Skip to #9)*

_____Yes, in certain circumstances. Please explain _____

_____*(Continue)*

8. If your answer to item 7 is "Yes" or "Yes, in certain circumstances", please identify the specific rifles your publication recommends **against** using for **high-power rifle competition**.

<u>Make</u>   <u>Model</u>   <u>Caliber</u>

_____

_____

_____

9. Have you or any other author who contributes to your publication written any articles since 1989 concerning the use of semiautomatic rifles and their suitability for use in hunting or organized competitive shooting? *(Exclude Letters to the Editor.)*

_____Yes *(Continue)*   _____No *(You are finished with the survey. Thank you.)*

10. If your answer to item 9 is "Yes", please submit a copy of the applicable article(s). Any material you are able to provide will be very beneficial to our study. Please indicate the publication, issue date and page for each article.

An agency may not conduct or sponsor, and a person is not required to respond to, the collection of information unless it displays a currently valid OMB control number.

Comments:

2. If your answer to item 1 is "Yes," please identify the specific centerfire rifles you recommend:

    (8)   Anything except Uzis.

    (9)   All study rifles except Uzi.

    (12) See attached articles.

3. Please explain circumstances to question 3: Does your publication recommend against the use of any semiautomatic rifles whose design is based on the AK 47, FN-FAL, HK91, HK93, SIG 550-1, or Uzi for use in hunting medium game (for example, turkey) or larger game (for example, deer)?

    (12) When the caliber is inappropriate or illegal for the specific game species.

4. Other rifle make recommendations in response to question 4: If your answer to item 3 is "Yes" or "Yes, in certain circumstances," please identify the specific rifles that you recommend against using for hunting medium game (for example, turkey) or larger game (for example, deer)?

    (12) See attached articles.

The following two items are for the responses to question 6: If your answer to item 5 is "Yes," please identify the specific centerfire semiautomatic rifles you recommend:

Model

    (5)   Springfield M1A and Colt AR-15.

Caliber

    (5)   7.62m (M1A) and .223 (Colt).

The following items are for questions 9 and 10 on articles written and the submission of these articles with the survey.

Article 1

    (8)   No articles enclosed.

    (9)   Semiautomatic Takes Tubb to HP Title.

    (10) No articles attached.

Article 2

    (9)   AR-15 Spaceguns Invading Match.

DEPARTMENT OF THE TREASURY
BUREAU OF ALCOHOL, TOBACCO AND FIREARMS
WASHINGTON, D.C. 20226

DEC 10 1997

O:F:S:DMS
3310

Dear Sir or Madam:

On November 14, 1997, the President and the Secretary
of the Treasury decided to conduct a review to
determine whether modified semiautomatic assault rifles
are properly importable under Federal law. Under
18 U.S.C. section 925(d)(3), firearms may be imported
into the United States only if they are determined to
be of a type generally recognized as particularly
suitable for or readily adaptable to sporting purposes.
The firearms in question are semiautomatic rifles based
on the AK47, FN-FAL, HK91, HK93, SIG SG550-1, and Uzi
designs.

As part of the review, the Bureau of Alcohol, Tobacco
and Firearms (ATF) is interested in receiving
information that shows whether any or all of the above
types of semiautomatic rifles are particularly suitable
for or readily adaptable to hunting or organized
competitive target shooting. We are asking that your
organization voluntarily complete the enclosed survey
to assist us in gathering this information. We
anticipate that the survey will take approximately
15 minutes to complete.

Responses must be received no later than 30 days
following the date of this letter; those received after
that date cannot be included in the review. Responses
should be forwarded to the Bureau of Alcohol, Tobacco
and Firearms, Department FG, P.O. Box 50860,
Washington, DC 20091. We appreciate any information
you care to provide.

Sincerely yours,

John W. Magaw
Director

Enclosure

ATF SURVEY OF STATE FISH AND GAME COMMISSIONS
FOR RIFLE USAGE
Page 1 of 2

State:_____

1. Do the laws in your state place any prohibitions or restrictions (other than seasonal) on the use of **high-power** rifles for **hunting medium game (for example, turkey) or larger game (for example, deer)?**

_____Yes *(Continue)* _____No *(Skip to #2)*

1a. If "Yes", please cite law(s) and briefly describe the restrictions.

_____

_____

_____

_____

_____

2. Do the laws in your state place any prohibitions or restrictions (other than seasonal) on the use of **semiautomatic** rifles for **hunting medium game (for example, turkey) or larger game (for example, deer)?**

_____Yes *(Continue)* _____No *(Skip to #3)*

2a. If "Yes", please cite law(s) and briefly describe the restrictions.

_____

_____

_____

_____

_____

An agency may not conduct or sponsor, and a person is not required to respond to, the collection of information unless it displays a currently valid OMB control number.

OMB No. 1512-0542

_____*(Continue)*

3. What, if any, is the minimum caliber or cartridge dimensions that may be used for **hunting medium game (for example, turkey) or larger game (for example, deer)?**

Caliber: _____ **OR** Dimensions:_____

_____There is no minimum.

4. Does your commission or state collect any data on the types of rifles used in your state for **hunting medium game (for example, turkey) or larger game (for example, deer)?**

_____Yes *(Continue)*   _____No *(You are finished with the survey. Thank you.)*

> **4a.** If "Yes", please provide hard copies of any such available data for the past two hunting seasons of 1995 and 1996. Any data that you provide will be most beneficial to our study.
>
> If you would like us to contact you regarding the data, please provide your name and phone number.
>
> Name:_____Phone:_____

Survey Fish and Game Commissions for Rifle Usage

STATE	Restrictions		Minimum Caliber or Cartridge		
	Q1	Q2	Q3	Q4	Q5
	HiPwr	Semiauto	Minimum Caliber	Minimum Cartridge	Collect Data
Alabama	Yes	Yes	Any center fire rifle	None	No
Alaska	Yes	No	No Centerfire for big game		No
Arizona	No	Yes	.22 mag or larger		No
Arkansas	Yes	No	None	None	No
California	No	No	See Question 1a	See Question 1a	No
Colorado	Yes	Yes	0.24		No
Connecticut	Yes	Yes			
Delaware	Yes	Yes			
Florida	Yes	Yes	No rimfire for deer		No
Georgia	Yes	No	.22 Centerfire or larger		No
Hawaii	No	No			
Idaho	Yes	Yes	.22 rimfire		No
Illinois	Yes	Yes	None	None	No
Indiana	Yes	Yes	None		No
Iowa	Yes	Yes	not provided		No
Kansas	Yes	Yes	.23 caliber or larger		No
Kentucky	No	No			
Louisiana	Yes	No	.22 Centerfire		No
Maine	Yes	No	.22 mag or larger		No
Maryland	Yes	Yes			
Massachusetts	Yes	No	None	None	Yes
Michigan	Yes	Yes	.23 or larger		No
Minnesota	Yes	No	0.23	1.285"	No
Mississippi	Yes	No	None	None	No
Missouri	Yes	Yes	None	None	No
Montana	No	No	None		No
Nebraska	No	No			
Nevada	No	No			No
New Hampshire	Yes	Yes		above .22 rimfire	No
New Jersey	Yes	Yes	None	None	No
New Mexico	Yes	No	.24 centerfire or larger		No
New York	Yes	Yes	Must be centerfire		No
North Carolina	Yes	No	None	None	No
North Dakota	Yes	Yes	.22 Centerfire or larger		No
Ohio	Yes	No	None	None	No
Oklahoma	Yes	Yes	.22 magnum		No
Oregon	Yes	Yes	.22 or .24 or larger		No
Pennsylvania	Yes	Yes	None	None	No
Rhode Island	Yes	Yes		.229 maximum	No
South Carolina	Yes	No	Must be larger than .22		No
South Dakota	Yes	No	None	None	No
Tennessee	Yes	Yes	.24 or larger caliber		No
Texas	Yes	No	None	None	No
Utah	Yes	No		None	No
Vermont	Yes	No			No
Virginia	Yes	Yes	.23 caliber for deer		No
Washington	Yes	Yes	.240 or larger for coyote		No
West Virginia	No	No		Any centerfire	No
Wisconsin	Yes	No	.22 caliber or larger		No
Wyoming	Yes	No		23/100 bullet dia.	No

Restrictions for High Powered Rifles

1a. Please cite law(s) and briefly describe the restrictions.

Alabama
(19) No automatic weapons, no silenced weapons.

Alaska
(23) Bison hunters must use a caliber capable of firing a 200-grain bullet having 2,000 pounds of energy at 100 yards.

Arkansas
(11) No rifles for turkey.

California
(22) Centerfire for big game, 10 gauge or smaller for resident small game.

Colorado
(10) Semiautomatic rifle may not hold more than 6 rounds.

Connecticut
(39) Shotgun only on public lands. Can use any type of rifle on private land.

Delaware
(40) No rifles - shotguns/muzzle loaders only.

Florida
(25) Machine guns and silencers not permitted for any hunting.

Georgia
(29) No hi-power rifles allowed for turkey hunting.

Hawaii
(49) Must have discharge of 1200 foot pounds.

Idaho
(30) No hi-power rifles allowed for hunting turkey.

Illinois
(12) Turkey or deer may not be hunted with rifle. Deer may not be hunted with muzzle loading rifle. No restriction on rifles for coyote, fox, and woodchuck, etc.

Indiana
(34) No hi-power rifles allowed for deer or turkey hunting. Limited restrictions for specified areas.

Iowa
(26) Cannot use rifles for turkey or deer, only shotgun or bow and arrow. No difference if public or private lands. For coyote or fox, there is no restriction on rifles, magazine size, or caliber.

Kansas
(33) Must use ammunition specifically designed for hunting.

Louisiana
(6) No rifles for turkey hunting. Rifles for deer hunting must be no smaller than .22 centerfire.

Maine
(32) No hi-power rifles for turkey and water fowl. Some limited restrictions for specific areas.

Maryland
(42) Some restrictions based on county. They are allowed in western and southern Maryland. Shotguns only in and around Baltimore and Washington, D.C.

Massachusetts
(14) Rifles not permitted for hunting deer and turkey.

Michigan
(27) No turkey hunting with hi-power rifle. No night hunting with hi-power rifle. Deer hunting with hi-power rifle allowed only in lower southern peninsula. Limited restrictions for specific areas.

Minnesota
(13) Caliber must be at least .23. Ammunition must have a case length of at least 1.285". .30 caliber M1 carbine cartridge may not be used.

Mississippi
(15) Restricts turkey hunting to shotguns. However quadriplegics may hunt turkey with a rifle.

Missouri
(5) Rifles not permitted for turkey. Self loading firearms for deer may not have a combined magazine + chamber capacity of more than 11 cartridges.

Nebraska
(43) Allowed and frequently used, but magazine capacity maximum is six rounds.

Nevada
(1) Answer to #3 refers to NAS 501.150 and NAS 503.142. Not for turkey.

New Hampshire
(7) Magazine capacity no more than 5 rounds. Prohibits full metal jacket bullets for hunting. Prohibits deer hunting with rifles in certain towns.

New Jersey
(17) No rifles.

New Mexico
(31) No hi-power rifles allowed for hunting turkey.

New York
(24) No semiautomatics with a magazine capacity of greater than 6 rounds; machineguns and silencers not permitted for any hunting. Limited restrictions for specific areas.

North Carolina
(20) Centerfire rifles not permitted for turkey hunting.

North Dakota
(28) No hi-power rifles for turkey hunting.

Ohio
(3) Prohibits high power rifles for turkey, deer and migratory birds. High power rifles can be used on all other legal game animals.

Oklahoma
(8) Centerfire rifles only for large game. Magazines for .22 centerfire rifles may not hold more than 7 rounds.

Oregon
(2) OAR 635-65-700(1) must be .24 caliber or larger center fire rifle, no full automatic; OAR 635-65-700(2) hunters shall only use centerfire rifle .22 caliber; OAR-65-700(5) no military or full jacket bullets in original or altered form. Limited restrictions for specific areas.

Pennsylvania
(16) Rifles not permitted in Philadelphia & Pittsburgh areas.

Rhode Island
(44) .22 center fire during the summer for woodchucks.

South Carolina
(18) No rifle for turkey, rifle for deer must be larger than .22 caliber

South Dakota
(50) Magazine not more than five rounds.

Tennessee
(37) No hi-power rifles allowed for turkey hunting.

Texas
(21) Rimfire ammunition not permitted for hunting deer, antelope, and bighorn sheep; machine guns and silencers not permitted for hunting any game animals.

Utah
(9) No rifles for turkey hunting.

Vermont
(47) Turkey size less than 10 gauge. Deer/moose/beer, no restriction on caliber.

Virginia
(48) 23 caliber or larger for deer and bear. No restrictions for turkey. No magazine restrictions, shotgun limited to 3 shells. Restrictions vary from county to county - approximately 90 different rifle restrictions in the State of Virginia based on the county restrictions. Sawed-off firearms are illegal to own unless with a permit, if barrel less than 16 inches for rifle, and 18 inches for shotgun.

Washington
(46) Hunting turkey limited to shotguns. Small game limited to shotguns.

Wisconsin

(36) No .22 rimfire rifles for deer hunting.

Wyoming

(4) Big game and trophy animals, firearm must have a bore diameter of at least 23/100 of an inch.

Restrictions for Semiautomatic Rifles

2a. Please cite law(s) and briefly describe the restrictions.

Alabama

(19) Turkey may not be hunted with a centerfire rifle or rimfire rifle. Semiautomatic rifles of proper caliber are legal for all types of hunting. No restrictions on magazine capacity, except wildlife management areas where centerfire rifles are restricted to 10 round max.

Arizona

(38) Magazine cannot hold more than 5 rounds.

Colorado

(10) Semiautomatic rifle may not hold more than 6 rounds.

Connecticut

(39) Shotgun only on public lands. Any type of rifle can be used on private land.

Delaware

(40) No rifles - shotguns/muzzle loaders only.

Florida

(25) No semiautomatic centerfire rifles having a magazine capacity greater than 5 rounds.

Idaho

(30) No hi-power rifles (including semiautomatic) allowed for turkey hunting.

Illinois

(12) See #1.

Indiana

(34) No hi-power rifles allowed for turkey hunting.

Iowa

(26) Cannot use rifles for turkey or deer, only shotgun or bow and arrow. No difference in public or private land. For coyote or fox, there is no restriction on rifle, magazine size, or caliber.

Kansas

(33) Must use ammunition specifically designed for hunting.

Maryland
(42) Some restrictions. Based on county. Shotguns only in and around Baltimore and Washington, D.C.

Michigan
(27) Unlawful to hunt with semiautomatic rifles capable of holding more than 6 rounds in magazine and barrel. Rimfire (.22 cal) rifles excluded from restrictions.

Missouri
(5) Combined magazine + chamber capacity may not be more than 11 cartridges.

New Hampshire
(7) Turkey may not be hunted with rifles. Rifles may not have magazine capacity of more than 5 cartridges.

New Jersey
(17) No rifles.

New York
(24) No semiautomatics with a magazine capacity of greater than 6 rounds.

North Dakota
(28) No hi-power rifles (including semiautomatics) may be used for hunting turkey.

Oklahoma
(8) See #1.

Oregon
(2) OAR 635-65-700(1) and (2) limits magazine capacity to no more than 5 cartridges.

Pennsylvania
(16) Semiautomatic rifles are not lawful for hunting in Pennsylvania.

Rhode Island
(44) Cannot use semiautomatic during the winter, only during the summer months for woodchucks (during daylight from April 1 to September 30).

Tennessee
(37) No hi-power rifles, including semiautomatics, allowed for turkey hunting.

Vermont
(47) Semiautomatic 5 rounds or less.

Virginia
(48) Semiautomatics are legal wherever rifles can be used. 23 caliber or larger for deer and bear. No restrictions for turkey. No magazine restrictions, shotgun limited to 3 shells. Restrictions vary from county to county - approximately 90 different rifle restrictions in the State of Virginia based on the county restrictions. Sawed-off firearms are illegal to own unless with a permit, if barrel less than 16 inches for rifle, and 18 inches for shotgun. Striker 12 - drums holds 12 or more rounds and is illegal.

Washington
(46) Cannot use fully automatic for hunting.

West Virginia
(45) Cannot use fully automatic firearms for hunting.

Comments Provided by Law Enforcement Agencies

(1)   No research.

(2)   No research.

(3)   NOBLE and others forwarded information to a U.S. Senator on circumstances concerning police officers killed or injured by these weapons. No data was provided.

(4)   No research.

(7)   The organization stated: "Most of the data available on guns and crime does not provide the detail needed to identify the types of guns listed. . . . We have conducted several surveys that refer to assault rifles generically, including the Survey of Inmates in State Correctional Facilities 1991, Survey of Inmates in Local Jails 1995, and the Survey of Adults on Probation 1995. The data on assault weapons has not been analyzed in the recently released Survey of Adults on Probation 1995 or in the yet to be released Survey of Inmates in Local Jails 1995.

"Our report Guns Used in Crime includes the results of an analysis of the stolen data from the FBI's National Crime Information Center database. Our analysis was limited to general categories of guns and calibers of handguns. The recent evaluation of the assault weapons ban funded by the National Institute of Justice analyzed a more recent set of the same data with an emphasis on assault weapons. The results of this evaluation were reported in Impact Evaluation of the Public Safety and Recreational Firearms Use Protection Act of 1994."

"BJS [Bureau of Justice Statistics] supports the Firearms Research Information System (FARIS). . . . This database contains firearms-related information from surveys, research, evaluations, and statistical reports. . . . We queried this database for any research on assault weapons. The results of the query include both the reports listed above, as well as several others. Please note that in BJS's report Guns Used in Crime refers to the report Assault Weapons and Homicide in New York City prepared by one of our grantees. While the data are from 1993, the report provides interesting insights into the use of assault weapons and homicide. Another source of data on assault weapons and crime is the FBI's Law Enforcement Officers Killed and Assaulted series, which records the type of gun used in killings of police officers. Several of the reports listed in the FARIS query used these data, including Cop Killers: Assault Weapons Attacks on America's Police, and Cops Under Fire: Law Enforcement Officers Killed with Assault Weapons or Guns with High Capacity Magazines."

(9)   Guns in America: National Survey on Private Ownership and Use of Firearms (May 1997) states: The 1994 NSPOF (National Survey of Private Ownership of Firearms) estimates for the total number of privately owned firearms is 192 million: 65 million handguns, 70 million rifles, 49 million shotguns, and 8 million other long guns.

DEPARTMENT OF THE TREASURY
BUREAU OF ALCOHOL, TOBACCO AND FIREARMS
WASHINGTON, D.C. 20226

DIRECTOR

DEC 1 0 1997

O:F:S:DMS
3310

Dear Sir or Madam:

On November 14, 1997, the President and the Secretary of the Treasury decided to conduct a review to determine whether modified semiautomatic assault rifles are properly importable under Federal law. Under 18 U.S.C. section 925(d)(3), firearms may be imported into the United States only if they are determined to be of a type generally recognized as particularly suitable for or readily adaptable to sporting purposes. The firearms in question are semiautomatic rifles based on the AK47, FN-FAL, HK91, HK93, SIG SG550-1, and Uzi designs.

As part of the review, the Bureau of Alcohol, Tobacco and Firearms (ATF) is interested in receiving information that shows whether any or all of the above types of semiautomatic rifles are particularly suitable for or readily adaptable to hunting or organized competitive target shooting.

Although ATF is not required to seek public comment on this study, the agency would appreciate any factual, relevant information concerning the sporting use suitability of the rifles in question.

Your voluntary response must be received no later than 30 days from the date of this letter; those received after that date cannot be included in the review. Please forward your responses to the Bureau of Alcohol, Tobacco and Firearms, Department TA, P.O. Box 50860, Washington, DC 20091.

Sincerely yours,

John W. Magaw

John W. Magaw
Director

(12) The respondent felt that definitions and usage should be subject to rulemaking. The respondent stated that limits on "sporting" use do not take into account firearms technology and its derivative uses among millions of disparate consumers. Millions of gun owners currently engage in informal target competition.

The respondent stated that the firearms are suitable for sporting purposes and that ATF's practice of making "ad hoc" revisions to import criteria disrupts legitimate commerce. The respondent recommends that all changes to criteria should be subject to rulemaking.

(19) The respondent submitted a brochure and a statement supported by seven letters from FFL's who sell the SLR-95 and 97 and ROMAK 1 and 2. The respondent and all the supporting letters attest to the suitability of these guns for hunting because (1) they are excellent for deer or varmint hunting; (2) they are used by many for target shooting; (3) their ammunition is readily available and affordable; and (4) they are excellent for young/new hunters because of low recoil, an inexpensive purchase price, durability, and light weight, as well as being designed only for semiautomatic fire.

(20) One respondent submitted results of its independently conducted survey, which consisted of 30 questions. The results of the survey suggest that 36 percent of those queried actually use AK47-type rifles for hunting or competition, 38 percent use L1A1-type rifles for hunting or competition, and 38 percent use G3-type rifles for hunting or competition. Other uses include home defense, noncompetitive target shooting, and plinking. Of those queried who do not currently own these types of rifles, 35 percent would use AK-type rifles for hunting or competition, 36 percent would use L1A1-type rifles for hunting or competition, and 37 percent would use G3-type rifles for hunting or competition.

(22) The respondent claims that the majority of the study rifles' length and calibers can be used only for sporting purposes. The respondent asserts that the only technical detail remaining after the 1989 decision that is similar to a military rifle is the locking system. After 1989, the imported rifles have no physical features of military assault rifles. All have features which can be found on any semiautomatic sporting/hunting rifle.

However, the respondent writes that the Uzi-type carbines are "not suitable for any kind of sporting events other than law enforcement and military competitions because the caliber and locking system do not allow precise shooting over long distances."

(23) One respondent, who imports the SAR-8 and SAR-4800 that are chambered for .308 Winchester ammunition, states that neither rifle possesses any of the characteristics of either the 1989 determination or the 1994 law. The respondent states that both are permitted in match rifle and other competitions. The respondent states that only two questions should be considered to determine hunting suitability of a rifle: Whether the caliber is adequate to take one or more game species and whether the gun is safe and reliable. The respondent states that there is no factual or legal basis to conclude that the rifles are not "particularly suitable" for sporting purposes.

(24) The respondent writes: "The particular firearms differ from other guns that are universally acceptable only in cosmetic ways. There is no functional difference between semiautomatic firearms based on the external features that have been keyed on in an attempt to implement the import restrictions of the 1994 Crime Bill. As further attempts to differentiate functionally identical firearms by these features for the purposes of culling out those that might be politically suitable for an administrative import ban is wrong."

(25) The respondent writes that the SLG95 was developed exclusively for hunting and competitive shooting. The respondent points out that it is capable of single firing only and cannot be reassembled for use as an automatic weapon. It is made for endurance and accuracy to 300 meters.

(26) The respondent recommends AK47 variants specifically, but believes all study rifles are suitable or adaptable for sporting. The respondent states that a Galil-chambered .308/.223 with a two-position rear sight, adjustable front sight, or scope mount channel, are reliable, durable, accurate, and suitable for hunting and organized competitive shooting. The respondent states that the Uzi, which chambers 9mm and 40 S&W, two-position rear sight, and an adjustable front sight is suitable for organized competitive target shooting.

(27) The respondent states that the SIG-SG550-1, in its original configuration, never possessed assault rifle features. The respondent states that is was built as a semiautomatic, not a fully automatic that was converted or modified to semiautomatic. It does have protruding pistol grip, and its ergonomics are geared toward its original design of goal-precision shooting. The respondent says that the name "Sniper" was a marketing decision, and it is extremely popular in .223 competitions. Its price isolates the gun to the competitor/collector.

2

(28)   Letters from H&K users were submitted in support of their
        continued importation and use as sporting arms. Specifically,
        the SR9 and PSG1 were said to be clearly suitable and utilized
        daily for hunting and target shooting. The respondent states
        that sport is defined as "an active pastime, diversion,
        recreation" and that the use of these is all the justification
        needed to allow their importation. The PSG1 has been imported
        since 1974, and the SR9 since 1990. The semiautomatic feature
        dates to turn of the century.

        The respondent states that the cost would dissuade criminals from
        using them. The respondent refers to ATF's reports "Crime Gun
        Analysis (17 Communities)" and "Trace Reports 1993-1996" to show
        that the H&K SR9 and PSG1 are not used in crime. In the 4-year
        period covered by the reports, not one was traced.

(29)   The respondent faults the 1989 report both for not sufficiently
        addressing the issue of ready adaptability, as well as for the
        limited definition of sporting purposes. The respondent states
        that sport is defined as "that which diverts, and makes mirth;
        pastime, diversion." The respondent says that the NRA sponsors
        many matches, and personally attests to the FN-FAL and HK91 as
        being perfectly suitable for such matches. The respondent states
        that the rifles are also used for hunting deer, rabbits, and
        varmints. Further, the respondent remarks that the use of these
        rifles in crime is minuscule.

Importer/Individual Letters

On January 15, 1998, the study group received a second submission from
Heckler and Koch, dated January 14, 1998. It transmitted 69 letters
from individuals who appeared to be answering an advertisement placed
in Shotgun News by Heckler and Koch. The study group obtained a copy
of the advertisement, which requested that past and current owners of
certain H&K rifles provide written accounts of how they use or used
these firearms. The advertisement stated that the firearms in
question, the SR9 and the PSG1, were used for sporting purposes such as
hunting, target shooting, competition, collecting, and informal
plinking. The advertisement also referred to the 120-day study and the
temporary ban on importation, indicating that certain firearms may be
banned in the future.

Synopses of Letters:

1.   The writer used his SR9 to hunt deer (photo included).

2.   The writer used his SR9 to hunt deer (photo included).

3.   The writer used his SR9 for informal target shooting and plinking.

4.   The writer used his SR9 for target practice and recreation.

5.   The writer (a police officer) used SR9 to hunt. Said that it's too
      heavy and expensive for criminals.

3

6.   The writer used his SR9 for competition.

7.   The writer used H&K rifles such as these around the farm to control wild dog packs.

8.   The writer used his SR9 to hunt deer.

9.   The writer used his SR9 to hunt, participate in target practice, and compete.

10. The writer used his H&K rifles for informal target shooting.

11. The writer used his SR9 to hunt elk because it's rugged, and to shoot targets.

12. The writer used his SR9 to target practice.

13. The writer used his HK91 to hunt varmints and compete in military rifle matches.

14. The writer does not use the firearms but is familiar with their use for target shooting, hunting, and competition.

15. The writer uses HK firearms for DCM marksmanship competition.

16. The writer used his HK93 for 100-yard club matches and NRA-high power rifle matches.

17. The writer does not own the firearms but enjoys shooting sports and collecting.

18. The writer used his HK91 to hunt deer, boar, and mountain goat and in high-power match competitions.

19. The writer used his SR9 to shoot targets and for competitions.

20. The writer used his HK91 to shoot varmints, hunt small and big game, and shoot long-range silhouettes.

21. The writer used his SR8 to hunt deer, target shoot, and plink.

22. The writer used his HK93 to shoot in club competitions.

23. The writer used his SR9 to shoot targets because the recoil does not impact his arthritis.

24. The writer (a police officer) does not own the firearm but never sees HKs used in crime.

25. The writer used his HKs for target shooting, competition, and collection.

26. The writer does not own the firearms but likes recreational target shooting.

27. Writer does not own the firearms but states, "Don't ban."

4

28. The writer used his SR9 for hunting deer, varmints, and groundhogs; for target shooting; and for occasional competitions.

29. The writer used his SR9 to hunt deer because it's accurate, rugged, and reliable.

30. The writer used his SR9 to hunt deer and elk.

31. The writer used his SR9 to target shoot.

32. The writer used his SR9 to hunt deer and target shoot.

33. The writer used his HK91 to shoot military rifle 100-yard competitions.

34. The writer used his SR9 for hunting varmints and coyotes, for target shooting, and for competitions.

35. The writer used his SR9 to hunt deer and target shoot.

36. The writer (a former FBI employee) used his SR9 for hunting varmints and for precision and target shooting.

37. The writer used his HK for target shooting and competition.

38. The writer used his SR9 for informal target shooting and plinking and his HK91 for bowling pin matches, high-power rifle competitions, informal target shooting, and plinking.

39. The writer used his SR9 to plink and shoot targets, saying it's too heavy for hunting.

40. The writer has an HK91 as part of his military collection and indicates it may be used for hunting.

41. The writer used his SR9 to target shoot.

42. The writer used his SR9 to hunt deer and target shoot.

43. The writer does not own the firearms but says, "Don't ban."

44. The writer used his SR9 and HK93 for hunting deer, for target shooting, and for home defense.

45. The writer states, "Don't ban."

46. Writer states, "Don't ban."

47. Writer states, "Don't ban."

48. The writer owns an SR9; no use was reported.

49. Writer used his SR9 to compete in club matches and "backyard competitions."

50. The writer used his HK to hunt boar and antelope.

51. The writer states, "Don't ban."

52. The writer (a police officer) does not own the firearms but states that the are not used by criminals.

53. The writer used his HK91 to hunt deer.

54. The writer (a police trainer) says that the PSG1 is used for police sniping and competitive shooting because it's accurate. He says that it's too heavy to hunt with and has attached an article on the PSG1.

55. The writer used her two PSG1s for target shooting and fun.

56. The writer used his SR9 and PSG1 to hunt and target shoot.

57. The writer used his two PSG1s to hunt and target shoot.

58. The writer provides an opinion that the SR9 is used to hunt and target shoot.

59. The writer used his PSG1 for hunting deer and informal target shooting.

60. The writer used his PSG1 to target shoot and plink.

61. The writer states, "Don't ban."

62. The writer used his HK91 to target shoot.

63. The writer used his HK91 to target shoot.

64. The writer (a U.S. deputy marshall) used his SR9 to shoot at the range.

65. The writer used his SR9 to hunt deer and coyotes.

66. The writer used his SR9 to competitively target shoot.

67. The writer used his SR9 to hunt deer and bear.

68. The writer uses military-type rifles like these for predator control on the farm.

69. The writer used his SR9 to target shoot, plink, and compete in DCM matches.

Comments Provided by Interest Groups

(7)   Impact Evaluation of the Public Safety and Recreational Firearms Use Protection Act of 1994, Final Report. March 13, 1997.

(8)   Identical comments were received from five members of the JPFO. They are against any form of gun control or restriction regardless of the type of firearm. References are made comparing gun control to Nazi Germany.

(9)   The respondent contends that police/military-style competitions, "plinking," and informal target shooting should be considered sporting. Note: The narrative was provided in addition to survey that Century Arms put on the Internet.

The respondent questions ATF's definition of "sporting" purposes. The respondent contends that neither the Bill of Rights nor the Second Amendment places restrictions on firearms based on use.

(13)   Citing the 1989 report, the respondent states that the drafters of the report determined what should be acceptable sports, thus excluding "plinking."

The respondent states that appearance (e.g., military looking) is not a factor in determining firearms' suitability for sporting purposes. It is their function or action that should determine a gun's suitability. Over 50 percent of those engaged in Practical Rifle Shooting use Kalashnikov variants. Further, citing U.S. vs. Smith (1973), the "readily adaptable" determination would fit all these firearms.

(14)   The respondent states that the vast majority of competitive marksmen shoot either domestic or foreign service rifles. Only 2-3 participants at any of 12 matches fire bolt-action match rifles. If service rifles have been modified, they are permitted under NRA rule 3.3.1.

The respondent says that attempts to ban these rifles "is a joke."

(15)   The respondent states that these firearms are used by men and women alike throughout Nebraska. All of the named firearms are used a lot all over the State for hunting. The AK47 has the same basic power of a 30/30 Winchester. All of these firearms function the same as a Browning BAR or a Remington 7400. Because of their design features, they provide excellent performance.

(16)   The respondent states that the Bill of Rights does not show the second amendment connected to "sporting purposes." The respondent says that all of the firearms in question are "service rifles," all can be used in highpower rifle competition (some better than others), but under no circumstances should "sporting use" be used as a test to determine whether they can be sold to the American public. The respondent states that "sporting use" is a totally bogus question.

(17) The respondent's basic concern is that the scope of our survey is significantly too narrow (i.e., not responsive to the Presidential directive, too narrow to address the problem, and inadequate to the task). The respondent states, "We do not indicate that our determination will impact modifications made to skirt law. We rely on the opinions of the 'gun press.' At a minimum, the Bureau should deny importation of: any semiautomatic capable of accepting with a capacity of more than 10 rounds, and any semiautomatic rifle with a capacity to accept more rounds than permitted by the State with the lowest number of permitted rounds. Deny any semiautomatic that incorporates cosmetically altered 'rule-beating' characteristics. Deny any semiautomatic that can be converted by using parts available domestically to any of the 1994 banned guns/characteristics. Deny any semiautomatic manufactured by any entity controlled by a foreign government. OR manufactured by a foreign entity that also manufactures, assembles or exports assault-type weapons. Deny any semiautomatic that contains a part that is a material component of any assault type weapon made, assembled, or exported by the foreign entity which is the source of the firearm proposed to be imported."

"A material component of any assault type weapon, assembled or exported by the foreign entity, which is, the source of the firearms proposed to be imported. The gun press has fabricated 'sporting' events to justify these weapons. The manner in which we are proceeding is a serious disservice to the American people."

Attachments: <u>That Was Then, This is Now; Assault Weapons; Analysis, New Research, and Legislation; Assault Weapons and Accessories in America; and Cop Killers.</u> All authored by the Violence Policy Center.

(30) The respondent states, "At least for handguns, and among young adult purchasers who have a prior criminal history, the purchase of an assault-type firearm is an independent risk factor for later criminal activity on the part of the purchaser."

NOTE: The above study was for assault-type handguns used in criminal activity versus other handguns. The study involved only young adults, and caution should be used in extending these results to other adults and purchasers of rifles. However, the respondent states, it is plausible that findings for one class of firearms may pertain to another closely related class.

(31) <u>The 1996 National Survey of Fishing, Hunting and Wildlife-Associated Recreation.</u> The publication outlines 1996 expenditures for guide use and percentage of hunters using guides for both big game and small game hunting.

(32)   In a memo from the Center to Prevent Handgun Violence the sections
       are Legal Background, History of Bureau Application of the
       "Sporting Purposes" Test, The Modified Assault Rifles under Import
       Suspension Should Be Permanently Barred from Importation, [The
       Galils and Uzis Should Be Barred from Importation Because They Are
       Banned by the Federal Assault Weapon Statute, and All the Modified
       Assault Rifles Should Be Barred from Importation Because They Fail
       the Sporting Purposes Test]. The conclusion states: "The modified
       assault rifles currently under suspended permits should be
       permanently barred from importation because they do not meet the
       sporting purposes test for importation under the Gun Control Act of
       1968 and because certain of the rifles [Galils and Uzis] also are
       banned by the 1994 Federal assault weapon law."

Comments Provided by Individuals

(10) The respondent does not recommend the Uzi, but he highly recommends the others for small game and varmints. He feels that the calibers of these are not the caliber of choice for medium or large game; however, he believes that the SIG and H&K are the best-built semiautomatics available.

He can not and will not defend the Uzi, referring to it as a "piece of junk."

The respondent feels that because of their expense and their being hard to find, the study rifles (excluding the Uzi) would not be weapons of choice for illegal activities.

(11) The respondent questions ATF's definition of "sporting" and "organized shooting." He feels that ATF's definition is too narrow and based on "political pressure."

The respondent feels that the firearms are especially suitable for competitive shooting and hunting and that the restrictions on caliber and number of cartridges should be left to the individual States. He has shot competitively for 25 years.

(18) The respondent specifically recommends the MAK90 for hunting because its shorter length makes for easier movement through covered areas, it allows for quicker follow-up shots, its open sights allow one to come up upon a target more quickly, and it provides a quicker determination of whether a clear shot exists through the brush than with telescopic sighting.

(21) The respondent states that the second amendment discusses "arms," not "sporting arms." The respondent further states that taxpayer money was spent on this survey and ATF has an agenda. A gun's original intent (military) has nothing to do with how it is used now. "The solution to today's crime is much the same as it always has been, proper enforcement of existing laws, not the imposition of new freedom-restricting laws on honest people."

Information on Articles Reviewed

(1) Describes limited availability of Uzi Model B sporter with thumbhole stock.

(2) Describes rifle and makes political statement concerning 1989 ban.

(3) Describes Chinese copy of Uzi with thumbhole stock.

(4) Quality sporting firearms from Russia.

(5) Short descriptions of rifles and shotguns available. Lead-in paragraph mentions hunting. Does not specifically recommend any of the listed weapons for hunting.

(6) Geared to retail gun dealers, provides list of available products. States L1A1 Sporter is pinpoint accurate and powerful enough for most North American big game hunting.

(7) Discusses the use of the rifle for hunting bear, sheep, and coyotes. Describes accuracy and ruggedness. NOTE: The rifle is a pre-1989 ban assault rifle.

(8) Deals primarily with performance of the cartridge. Makes statement that AK 47-type rifle is adequate for deer hunting at woods ranges.

(9) Discusses gun ownership in the United States. Highlighted text (not by writers) includes the National Survey of Private Ownership of Firearms that was conducted by Chilton Research Services of Drexel Hill, Pennsylvania during November and December 1994: 70 million rifles are privately held, including 28 million semiautomatics.

(10) Discusses pre-1989 ban configuration. Describes use in hunting, and makes the statement that "in the appropriate calibers, the military style autoloaders can indeed make excellent rifles, and that their ugly configuration probably gives them better handling qualities than more conventional sporters as the military discovered a long time ago."

(15) Not article - letter from Editor of Gun World magazine discussing "sport" and various competitions. Note: Attached submitted by Century Arms.

(16) Letter addressed to "To Whom It May Concern" indicating HK91 (not mentioned but illustrated in photos) is suitable for hunting and accurate enough for competition. Note: Submitted by Century Arms.

(17) Describes a competition developed to test a hunter's skill. Does not mention any of the rifles at issue.

(18) Not on point - deals with AR 15.

(19) Describes function, makes political statement.

(20) Discusses function and disassembly of rifle.

(21) Not on point - deals with AR 15 rifle.

(22) Discusses competition started to show sporting use of rifles banned for sale in California. Unknown if weapons in study were banned in California in 1990.

(23) Not on point - deals with national matches.

(24) Not on point - deals with various surplus military rifles.

(25) Deals with 7.62x39mm ammunition as suitable for deer hunting and mentions the use in SKS rifles, which is a military style semiautomatic but not a part of the study.

(26) Not on point - deals with reloading.

(27) Not on point - deals with reloading.

(28) Not on point - deals with AR15 rifles in competition.

(29) Not on point - deals with the SKS rifle.

(30) Not on point - deals with national matches.

(31) Not on point - deals with national matches.

(32) Not on point - deals with national matches.

(33) Not on point - deals with national matches at Camp Perry.

(34) Not on point - deals with national matches at Camp Perry.

(35) Not on point - deals with 1989 national matches at Camp Perry.

(36) Not on point - deals with Browning BAR sporting semiautomatic rifles.

(38) Not on point - deals with AR15, mentions rifle in caliber 7.62 x 39.

(39) Not on point - deals with bullet types.

(40) Not on point - deals with reloading.

(41) Discusses tracking in snow. Rifles mentioned do not include any rifles in study.

(42) Deals with deer hunting in general.

(43) Deals with rifles for varmint hunting. Does not mention rifles in study.

(44) Not on point - deals with hunting pronghorn antelope.

(45) Deals with various deer rifles.

(46) Not on point - deals with two Browning rifles' recoil reducing system.

(47) Not on point - deals with bolt-action rifles.

(48) Not on point - deals with ammunition.

(49) Deals with modifications to AR15 trigger for target shooting.

(50) Not on point - deals with M1 Garand as a target rifle.

(51) Not on point - deals with reloading.

(52) Deals with impact of banning semiautomatic rifles would have on competitors at Camp Perry.

(53) Deals with economic impact in areas near Camp Perry if semiautomatic rifles banned. Reprint from <u>Akron Beacon Journal</u>.

(54) Deals with training new competitive shooters - mentions sporting use of assault rifles, i.e., AR15.

(55) Not on point - article about Nelson Shew.

(56) Not on point - deals with reloading.

(57) Not on point - deals with shooting the AR15.

(58) Not on point - article about AR15 as target rifle.

(59) Not on point - article about well known competitive shooter.

(67) Not on point - deals with reloading.

(68) Discusses semiautomatic versions of M14.

(69) Discusses gas operation.

(70) Discusses right adjustment on M1 and M1A rifles.

(71) Discusses M1A and AR15-type rifles modified to remove them from assault weapon definition, and their use in competition.

(72) Deals with AR15 type rifle.

(73) Not on point - deals with AR15.

(74) Not on point - deals with target rifle based on AR15/M16.

(75) Not on point - deals with SKS rifle.

(76) Not on point - deals with reloading 7.62x39mm cartridge.

(77) Not on point - deals with reloading. Mentions 7.62x39mm.

(78) Not on point - deals with ammunition performance.

(79) Deals with .223 Remington caliber ammunition as a hunting cartridge.

(80) Describes M1A (semiautomatic copy of M14) as a target rifle.

(81) Not on point - deals with bullet design.

(82) Not on point - deals with ammunition performance.

Information on Advertisements Reviewed

(11) Indicates rifles are rugged, reliable and accurate.

(12) Describes rifles, lists price.

(13) Sporting versions of AK 47 and FAL.

(14) Sporting version of AK 47, reliable, accurate.

(61) Catalog of ammunition - lists uses for 7.62x39mm ammunition.

(62) Catalog of ammunition - lists uses for 7.62x39mm ammunition.

(63) Catalog of ammunition - lists uses for 7.62x39mm ammunition.

(64) Catalog of ammunition - lists uses for 9mm ammunition.

(65) Catalog of ammunition - lists uses for 9mm ammunition.

(66) Catalog of ammunition - lists recommended uses for 9mm ammunition.

www.ingramcontent.com/pod-product-compliance
Lightning Source LLC
Chambersburg PA
CBHW081458170526
45166CB00008B/2472